Ultrasound for Material Characterization and Processing

Ultrasound for Material Characterization and Processing

Editor

Francesca Lionetto

MDPI • Basel • Beijing • Wuhan • Barcelona • Belgrade • Manchester • Tokyo • Cluj • Tianjin

Editor
Francesca Lionetto
Engineering for Innovation
University of Salento
Lecce
Italy

Editorial Office
MDPI
St. Alban-Anlage 66
4052 Basel, Switzerland

This is a reprint of articles from the Special Issue published online in the open access journal *Materials* (ISSN 1996-1944) (available at: www.mdpi.com/journal/materials/special_issues/Ultrasound_for_ Material_Characterization_Processing).

For citation purposes, cite each article independently as indicated on the article page online and as indicated below:

LastName, A.A.; LastName, B.B.; LastName, C.C. Article Title. *Journal Name* **Year**, *Volume Number*, Page Range.

ISBN 978-3-0365-1710-0 (Hbk)
ISBN 978-3-0365-1709-4 (PDF)

Contents

About the Editor

Francesca Lionetto

Francesca Lionetto is an Assistant Professor at the University of Salento in Italy. She has been a visiting scientist at the University of Nottingham (UK) and at the University of Kaiserslautern (Germany). She is the author of more than 130 scientific publications on international journals, conference proceedings, and book chapters. She is a member of the Editorial Board of two international peer-reviewed scientific journals and a reviewer for about 30 international journals. She is the author of two national patents and one submitted European patent. Her research activity is focused on polymer and polymer composite science and technology. Her research interests are i) polymer rheology; ii) curing kinetics of thermosetting matrices; iii) polymer composite processing and joining; iv) heat transfer modelling during composite manufacturing; v) polymer-based nanocomposites; vi) hybrid welding of dissimilar materials; and vii) ultrasonic wave propagation.

materials

MDPI

Editorial

Ultrasound for Material Characterization and Processing

Francesca Lionetto

Department of Engineering for Innovation, University of Salento, Via Monteroni, 73100 Lecce, Italy;
francesca.lionetto@unisalento.it

Citation: Lionetto, F. Ultrasound for
Material Characterization and
Processing. *Materials* **2021**, *14*, 3891.
https://doi.org/10.3390/ma14143891

Received: 5 July 2021
Accepted: 9 July 2021
Published: 12 July 2021

Publisher's Note: MDPI stays neutral
with regard to jurisdictional claims in
published maps and institutional affiliations.

Ultrasonic waves are nowadays used for multiple purposes in many different fields from the non-destructive inspection of materials to sonochemical synthesis of materials and welding. Usually, ultrasonic applications are divided into low intensity/high frequency ultrasound and high intensity/low frequency ultrasound. Low intensity ultrasound transmits energy through the medium in order to obtain the information about the medium or to convey information through the medium [1]. It is successfully used in non-destructive inspection, ultrasonic dynamic analysis, ultrasonic rheology, ultrasonic spectroscopy of materials, process monitoring, applications in civil engineering, aerospace and geological materials and structures and in the characterization of biological media [2]. Nowadays, it is an essential tool for assessing metals, plastics, aerospace composites, wood, concrete and cement [3]. High intensity ultrasound deliberately affects the propagation medium through the high local temperatures and pressures generated [4]. It is used in industrial processes such as welding, cleaning, emulsification, atomization, etc.; chemical reactions and reactor induced by ultrasonic waves; synthesis of organic and inorganic materials; microstructural effects; heat generation; accelerated material characterization by ultrasonic fatigue testing; food processing, environmental protection [5].

The Special Issue *Ultrasound for Material Characterization and Processing* collects eleven papers, one review and ten research papers, with the aim to present recent advances in ultrasonic wave propagation applied for the characterization or the processing of materials. Both fundamental science and applications of ultrasound in the field of material characterization and material processing have been gathered.

As regards low intensity ultrasound, Waldner et al. [6] in their review entitled *"Ultrasonic liquid penetration measurement in thin sheets—physical mechanisms and interpretation"*, give an overview on the three possible mechanisms for ultrasonic liquid penetration (ULP) measurements for the characterization of intrinsic properties of porous sheets. The review discusses the interpretation of ULP results, which are complex, as the ultrasound signal can be affected by several mechanisms. Then, it describes the individual physical mechanisms leading to different shapes of the ultrasound intensity curves, and provides measurements confirming that the measured ultrasound intensity is linked to the liquid penetration regime into the structure.

Acquaticci et al. [7] in the paper entitled *"Ultrasound Axicon: systematic approach to optimize focusing resolution through human skull bone"* propose a numerical approach for designing axicon lenses for many high-resolution applications, such as mapping or detection. The aim is to generate a focused ultrasound field by plane transducer using axicon lens, which can be suitable for focused ultrasound through human skull bone. The numerical simulations help to find the optimized parameters for the proposed system design. The sound field testing using hydrophone shows the performance of the proposed method by comparing the cases with and without a human skull phantom.

Lin et al. [8] in the paper entitled *"Effects of loading and boundary conditions on the performance of ultrasound compressional viscoelastography: a computational simulation study to guide experimental design"* present a finite element analysis of compressional viscoelastography. This ultrasonic imaging technique is used for measuring the viscoelastic properties of biomaterials and tissues and it is based on the analysis of the creep behavior of a material under an external mechanical compression. The authors demonstrate that loading conditions (the distribution of the applied compressional pressure on the surface of the sample) and boundary conditions (the fixation method used to stabilize the sample) can severely affect the measurement accuracy of the viscoelastic properties of materials.

Ryuzono et al. [9] in the paper entitled *"Topology optimization-based damage identification using visualized ultrasonic wave propagation"* present a new damage identification method in structures based on topology optimization in combination with visualized ultrasonic wave propagation. The proposed method is applied to an aluminum plate with an artificial crack and the estimated damage state and the sensitivity of the objective function were compared with numerical predictions.

Lionetto et al. [10] in the paper entitled *"Out-of-plane permeability evaluation of carbon fiber preforms by ultrasonic wave propagation"* present a novel experimental set-up, based on ultrasonic wave propagation, for the determination of the out-of-plane permeability of carbon fiber reinforcements, which is the dominant property in the infusion of large and flat panels with a high thickness. An accurate characterization of the reinforcement permeability is fundamental in order to estimate the optimum process parameters for manufacturing high-quality components. The experimental results, obtained in unsteady conditions, have been compared with those obtained in steady conditions by a traditional gravimetric method and validated by some analytical models. The work demonstrates the feasibility and potential of the proposed set-up for permeability measurements thanks to its non-invasive character and the one-side access.

In the area of structural mechanics, concrete structures and nondestructive testing Zima et al. [11] in the paper entitled *"Numerical study of concrete mesostructure effect on lamb wave propagation"* analyze the propagation of guided wave in concrete plates accounting for the heterogeneous structure of concrete. The influence of aggregate ratio and particle size on dispersion curves representing Lamb wave modes is analyzed by several concrete models with randomly distributed aggregates, generated by means of the Monte Carlo method. The study shows that Lamb wave propagates with different velocities in homogeneous and heterogeneous models and the difference increases with aggregate ratio and particle size. This result demonstrates the potential of the method for the monitoring of concrete structures.

Wang et al. [12] in the paper entitled *"Monitoring the Setting Process of Cementitious Materials Using Guided Waves in Thin Rods"* demonstrate the reliability of an ultrasonic method based on embedded guided waves to monitor the early-age properties of mortar and concrete samples through continuous attenuation monitoring. The technique is able to eliminate the effects of coarse aggregates, which makes it of great potential to be directly applied in situ to the fresh concrete. The method has the potential to be used when additives are added to concrete, thus changing its rheological properties.

Concerning the applications of high intensity ultrasound, Frederick et al. [13], in the paper entitled *"Disassembly study of ultrasonically welded thermoplastic composite joints via resistance heating"*, demonstrate a novel concept in the disassembly of ultrasonically welded thermoplastic composite joints. The recycling of thermoplastic composite joints is realized by resistance heating via a nanocomposite film located at the welded interface. This electrically conductive film, containing multi-walled carbon nanotubes, is able to promote joint disassembly through a combination of nanocomposite and matrix melting and weakened fiber–matrix interface.

Shan et al. [14], in the paper entitled *"Building of longitudinal ultrasonic assisted turning system and its cutting simulation study on bulk metallic glass"*, present an experimental-numerical study on a novel machining technology, longitudinal ultrasonic assisted turning. After presenting this technology, they establish a two-dimensional finite element model of the ultrasonic assisted turning, aimed at studying the effect of ultrasonic vibration on cutting force under different cutting speeds on bulk metallic glasses. The latter are a new kind of material, which are made by rapid condensation of alloy with excellent properties such as high strength, high hardness and corrosion resistance, but they are hard to be machined. The results demonstrate that longitudinal ultrasonic vibration can significantly reduce the average cutting force as well as the von Mises stress when the turning speed is below the critical turning speed.

Shi et al. [15], in the paper entitled *"Effects of ultrasonic bending vibration introduced by an L-shaped ultrasonic rod on the microstructure and properties of a 1060 aluminum alloy strip formed by twin-roll casting"*, present a new type of ultrasonic generator to assist the process of continuous casting and rolling of aluminum alloy. A novel L-shaped ultrasonic rod is designed to introduce an ultrasonic bending vibration into the aluminum melt. The ultrasonic bending vibration improves the microstructure of the 1060 aluminum alloy cast rolling strip, characterized by grain refinement, and consequently its properties.

Shi et al. [16], in the paper entitled *"Effect of ultrasonic bending vibration introduced by the L-shaped ultrasonic rod on solidification structure and segregation of large 2A14 ingots"*, propose a new design of L- shaped ceramic ultrasonic wave guide rod for the introduction of ultrasonic bending vibration during the solidification of the 2A14 aluminum alloy. The authors demonstrate the advantages in terms of grain refinement, deagglomeration of the secondary phase, and improving the ingot macrostructure.

Funding: This research received no external funding.

Acknowledgments: I would like to kindly acknowledge Jason Huang, Assistant Editor of Materials journal, for his support for the success of during all the steps of the Special Issue. I am also grateful to the entire staff of the Materials Editorial Office for the collaboration. Moreover, all the contributors and reviewers are greatly acknowledged for their excellent work.

Conflicts of Interest: The author declares no conflict of interest.

References

1. Lionetto, F.; Maffezzoli, A.; Ottenhof, M.A.; Farhat, I.A.; Mitchell, J.R. Ultrasonic investigation of wheat starch retrogradation. *J. Food Eng.* **2006**, *75*, 258–266. [CrossRef]
2. Espinoza-Gonzalez, C.; Avila-Orta, C.; Martinez-Colunga, G.; Lionetto, F.; Maffezzoli, A. A Measure of CNTs Dispersion in Polymers with Branched Molecular Architectures by UDMA. *IEEE Trans. Nanotechnol.* **2016**, *15*, 731–737. [CrossRef]
3. Lionetto, F.; Maffezzoli, A. Relaxations during the postcure of unsaturated polyester networks by ultrasonic wave propagation, dynamic mechanical analysis, and dielectric analysis. *J. Polym. Sci. Part B Polym. Phys.* **2005**, *43*, 596–602. [CrossRef]
4. Lionetto, F.; López-Muñoz, R.; Espinoza-González, C.; Mis-Fernández, R.; Rodríguez-Fernández, O.; Maffezzoli, A. A Study on Exfoliation of Expanded Graphite Stacks in Candelilla Wax. *Materials* **2019**, *12*, 2530. [CrossRef] [PubMed]
5. Dell'Anna, R.; Lionetto, F.; Montagna, F.; Maffezzoli, A. Lay-Up and Consolidation of a Composite Pipe by In Situ Ultrasonic Welding of a Thermoplastic Matrix Composite Tape. *Materials* **2018**, *11*, 786. [CrossRef] [PubMed]
6. Waldner, C.; Hirn, U. Ultrasonic liquid penetration measurement in thin sheets—Physical mechanisms and interpretation. *Materials* **2020**, *13*, 2754. [CrossRef]
7. Acquaticci, F.; Lew, S.E.; Gwirc, S.N. Ultrasound axicon: Systematic approach to optimize focusing resolution through human skull bone. *Materials* **2019**, *12*, 3433. [CrossRef]
8. Lin, C.-Y.; Chang, K.-V. Effects of Loading and Boundary Conditions on the Performance of Ultrasound Compressional Viscoelastography: A Computational Simulation Study to Guide Experimental Design. *Materials* **2021**, *14*, 2590. [CrossRef] [PubMed]
9. Ryuzono, K.; Yashiro, S.; Nagai, H.; Toyama, N. Topology optimization-based damage identification using visualized ultrasonic wave propagation. *Materials* **2020**, *13*, 33. [CrossRef]
10. Lionetto, F.; Montagna, F.; Maffezzoli, A. Out-Of-Plane Permeability Evaluation of Carbon Fiber Preforms by Ultrasonic Wave Propagation. *Materials* **2020**, *13*, 2684. [CrossRef] [PubMed]
11. Zima, B.; Rafael, K. Numerical Study of Concrete Mesostructure Effect on Lamb Wave Propagation. *Materials* **2020**, *13*, 2570. [CrossRef] [PubMed]
12. Wang, D.; Yu, G.; Liu, S.; Sheng, P. Monitoring the Setting Process of Cementitious Materials Using Guided Waves in Thin Rods. *Materials* **2021**, *14*, 566. [CrossRef] [PubMed]
13. Frederick, H.; Li, W.; Palardy, G. Disassembly Study of Ultrasonically Welded Thermoplastic Composite Joints via Resistance Heating. *Materials* **2021**, *14*, 2521. [CrossRef] [PubMed]
14. Shan, S.; Feng, P.; Zha, H.; Feng, F. Building of Longitudinal Ultrasonic Assisted Turning System and Its Cutting Simulation Study on Bulk Metallic Glass. *Materials* **2020**, *13*, 3131. [CrossRef] [PubMed]
15. Shi, C.; Fan, G.; Mao, X.; Mao, D. Effects of ultrasonic bending vibration introduced by an L-shaped ultrasonic rod on the microstructure and properties of a 1060 aluminum alloy strip formed by twin-roll casting. *Materials* **2020**, *13*, 2013. [CrossRef] [PubMed]
16. Shi, C.; Wu, Y.; Mao, D.; Fan, G. Effect of Ultrasonic Bending Vibration Introduced by the L-shaped Ultrasonic Rod on Solidification Structure and Segregation of Large 2A14 Ingots. *Materials* **2020**, *13*, 807. [CrossRef] [PubMed]

 *materials*

Review

Ultrasonic Liquid Penetration Measurement in Thin Sheets—Physical Mechanisms and Interpretation

Carina Waldner [1,2] and Ulrich Hirn [1,2,*]

1   Institute of Bioproducts and Paper Technology, TU Graz, Inffeldgasse 23, 8010 Graz, Austria; carina.waldner@tugraz.at
2   CD Laboratory for Fiber Swelling and Paper Performance, Inffeldgasse 23, 8010 Graz, Austria
*   Correspondence: ulrich.hirn@tugraz.at

Received: 30 April 2020; Accepted: 10 June 2020; Published: 17 June 2020

Abstract: Ultrasonic liquid penetration (ULP) measurements of porous sheets have been applied for a variety of purposes ranging from determining liquid absorption dynamics to surface characterization of substrates. Interpretation of ULP results, however, is complex as the ultrasound signal can be affected by several mechanisms: (1) air being replaced by the liquid in the substrate pores, (2) air bubbles forming during penetration, and (3) structural changes of the substrate due to swelling of the substrate material. Analyzing tailored liquids and substrates in combination with contact angle measurements we are demonstrating that the characteristic shape of the ULP measurement curves can be interpreted in terms of the regime of liquid uptake. A fast and direct decline of the curve corresponds to capillary penetration, the slope of the curve indicates the penetration speed. A slow decline after a previous maximum in the signal can be related to diffusive liquid transport and swelling of the substrate material.

Keywords: liquid penetration; ultrasound transmission; capillary penetration; porous sheets

1. Introduction

Ultrasound attenuation measurements during liquid penetration have originally been reported to be able to determine the degree of sizing in paper [1] and correlations to liquid absorption measurements (Bristow absorption test) have been found [2]. Since then, ultrasonic liquid penetration (ULP) measurements have been applied to study different aspects of interactions between liquids and porous, sheet like materials (paper, membranes, and thin films). Things investigated were for example network hydrophobization (sizing) [3–6], kinetics of liquid absorption (wetting and penetration) [7–10], evaluating water/grease barrier properties [11–14], as well as predicting print quality [15–18] and gluing strength [19]. The method has even been used to characterize the structure of the paper substrate, e.g., surface pores and roughness [20–23].

In these publications, a wide range of liquids and substrates were combined, ranging from all kinds of papers to membranes and films on the substrate side and from water, to alcohols, oils, glues, coating colors, inks, and all kinds of mixtures on the liquid side. There is hardly any limit to combinations of substrates and liquids that can be combined and investigated with ULP. The only requirement for a sample to be measurable seems to be that attenuation is not too high so that a signal can still be detected at the receiver. This is affected by a combination of factors like sample thickness, porosity, pore size, and other material properties like density and elasticity. Furthermore, the signal frequency is playing a large role. For samples measured with 2 MHz, thickness was between 40 μm and 2 mm, pore size was between 0.01 and 100 μm, and porosity was between 2% and 90%.

The measurement principle is shown in Figure 1a. The sample is fixed on a sample holder with two-sided adhesive tape. When the test is started, the sample holder is quickly immersed into the test liquid. As soon as the sample is completely immersed, a high frequency (typically 1 or 2 MHz), low energy ultrasound signal is transmitted through the sample in thickness direction and the changes in transmission are recorded over time. The signal is a pulse with an initial pulse length in the sender of about 1 µs. At the receiver, the pulse builds up over seven oscillations and declines over seven oscillations. Consequently, no difference compared to a continuous wave regarding the interference should be seen in practice [24].

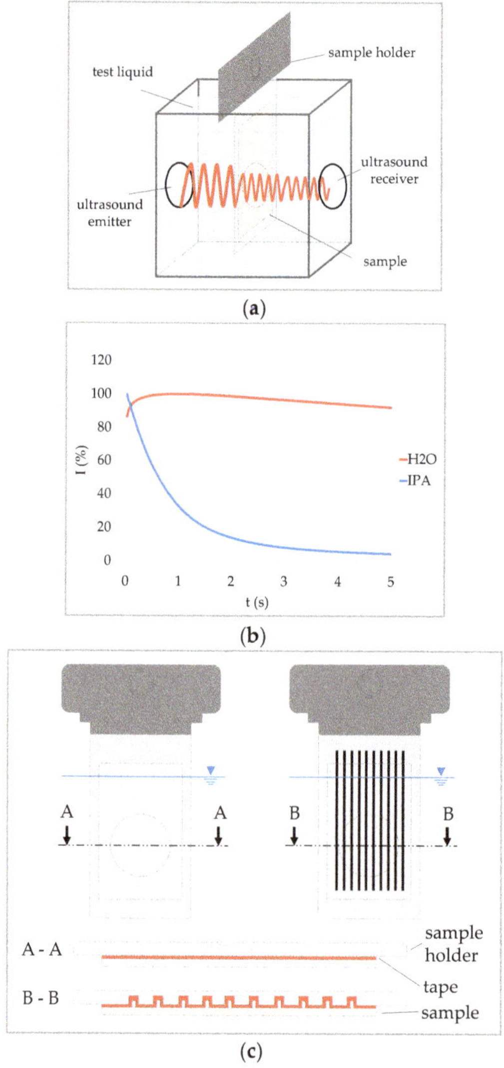

Figure 1. Ultrasonic liquid penetration (ULP) measurement: (**a**) measurement principle: the sample is immersed in the liquid; ultrasound is transmitted through the sample in thickness direction and the transmitted signal intensity I is recorded over time; (**b**) typical ULP curves of water and an isopropanol-water mixture (IPA) on sized copy paper on flat sample holder showing the relative change of the ultrasound intensity at the receiver; the two liquids lead to different curve shapes; and (**c**) different types of sample holders: smooth sample holder (left), sample holder with grooves (right; drawing not to scale).

Figure 1c shows two types of sample holders. On the flat sample holder (left) the sample backside is completely covered with the tape, while the sample is not in contact with the tape in the grooves of the grooved sample holder. Unless otherwise stated measurements discussed in this paper were made with the flat sample holder.

Figure 1b shows two typical ultrasound transmission curves of water and an isopropanol–water mixture on a sized copy paper on the flat sample holder. The curves show the ultrasound transmission intensity over time, as a percentage of the maximum ultrasound signal collected. Thus, only the changes of the signal during penetration affect the shape of the curve. In Figure 1b, the curve for water initially exhibits an increase in intensity, followed by a decline. The curve for isopropanol/water immediately declines rapidly. In order to be able to interpret these curves, it is necessary to understand mechanisms that influence the signal during liquid penetration. Several mechanisms are said to have an impact on the signal during liquid penetration (compare for instance [5,6,15,22]):

- Air being replaced by the liquid;
- Air bubbles forming during penetration;
- Structural changes of the substrate (paper) due to swelling of the fibers.

However, it is not very well understood how these processes affect the signal transmission, how the individual mechanisms are related to the shape of the measurement curves, and which of the mechanisms has the strongest impact. Without this understanding, interpretation of results is difficult. For instance, the time to the maximum intensity has been interpreted as the wetting time (e.g., [3,5,15]), but no correlation with wetting measured via contact angle measurements could be found [8]. Therefore, the aim of this paper is to clarify how these mechanisms influence ultrasound transmission, which of the mechanisms is dominant in certain circumstances and finally, to provide a guideline for interpretation of the result curves. To be able to improve the understanding of how the individual mechanisms influence ultrasound transmission during liquid penetration, first, fundamentals of ultrasound transmission and liquid penetration are discussed. Then we will discuss the individual physical mechanisms leading to different shapes of the ultrasound intensity curves, and provide measurements confirming that the measured ultrasound intensity is indeed linked to the liquid penetration regime into the structure. Finally, we describe different parameters describing liquid penetration in thin substrates obtained from ULP measurements.

2. Fundamentals of Ultrasound Transmission

On its way through the sample, the ultrasound wave is attenuated, i.e., the amplitude and ultrasound intensity decrease. The ultrasound intensity is defined as the energy flow rate transmitted through a unit area perpendicular to the propagation direction and is proportional to the square of the amplitude [25]. Attenuation is mainly caused by thermal and viscous losses, reflection occurring at phase boundaries, and scattering. Thermal and viscous losses have only a negligible impact in the measurement setup shown in Figure 1a [24]. Since the signal attenuation within a material is not relevant, attenuation is dominated by reflection and scattering at material interfaces.

2.1. Attenuation Due to Reflection

Attenuation of the ultrasound signal due to reflection at phase boundaries occurs if two adjacent media have differing wave impedances. The larger the wave impedance difference, the more reflection occurs. The wave impedance can be calculated for many materials with: [26]

$$Z = \rho \times c, \tag{1}$$

With Z being the wave impedance (kg/m^{2}s), ρ the material's density (kg/m^3), and c the speed of sound in the material (m/s). For the case of perpendicular incidence of an ultrasonic compressional wave the reflection coefficient R at the boundary is: [26]

$$R = \frac{Z_1 - Z_2}{Z_1 + Z_2} \qquad (2)$$

Z_1 and Z_2 being the wave impedances of the two media. Equation (2) shows that less reflection occurs if the wave impedances of the two media are similar. Table 1 shows the wave impedances for air, water, and pulp fibers in paper calculated from density and speed of sound. During the penetration of a paper sample with water, the ultrasound signal can be reflected at the boundaries between air and water, air and fibers, and water and fibers. From the wave impedances, the reflection coefficient at each type of boundary can be calculated (see Table 2).

Table 1. Wave impedances for air, water, and pulp fibers.

Medium	Speed of Sound (m/s)	Density (kg/m^3)	Wave Impedance (kg/m^{2}s)
air [1]	343	1.29	442.47
water [1]	1490	998	1,487,020
fiber [2]	1493	1500	2,239,548

[1] values at 20 °C taken from [27]; [2] values based on [28].

Table 2. Reflection coefficients for air, water, and fiber boundaries.

$R_{air,water}$	$R_{air,fiber}$	$R_{water,fiber}$
0.9988	0.9992	0.0408

The reflection coefficients show that at an air-fiber boundary most of the signal was reflected. The same is true for air–water boundaries. If the air within a paper sample is completely replaced by water, water–fiber boundaries are replacing the air-fiber boundaries. The reflection coefficient for water-fiber boundaries was much lower, yielding a better signal transmission. This means that in general the received ultrasound signal should increase when the air in the sample is replaced by the test liquid during penetration. Please note that the wave impedance value for fibers should not be understood to be valid for all paper fibers, since the speed of sound in the fibers largely depends on the manufacturing process (compare [29]).

The system of air, water, and pulp fibers is used as an example for a liquid penetrating into a porous material. Since solids and liquids generally have more similar wave impedances compared to gases [27], the considerations made above are also valid for other porous solids being penetrated by any kind of liquid. In conclusion, displacement of air by liquid in the substrate pores thus can only be responsible for an increase in ultrasound intensity over time, the curves however often show a (pronounced) decrease.

2.2. Attenuation Due to Scattering

Liquid penetrating into a porous substrate represents a very inhomogeneous system in which water, fibers/pores, and air bubbles are distributed irregularly. In such an inhomogeneous system, scattering of the ultrasound signal influences the received signal [27]. Since the transmitted ultrasound intensity is normalized in the results, only factors that could change the scattering behavior during penetration are relevant.

As argued above, the air being replaced by the liquid should generally lead to a signal increase due to the lower reflection coefficient of a solid–liquid boundary. However, if the air cannot escape during penetration, air bubbles could be entrapped within the porous substrate and constitute additional scattering centers, ultimately leading to a signal decrease. Daun [24] showed how bubbles of different sizes impact scattering (see Figure 2). The scattering cross section thereby refers to the ratio of the total scattered power and the intensity of the incoming signal. The total scattered power is obtained by integrating the scattered intensity in each direction over the surface of a sphere. Taking resonance of bubbles into account, small bubbles of a critical size can increase scattering effects by several orders of magnitude. Figure 2 also shows that resonance effects depend on the frequency of the ultrasound signal. For 1 MHz the scattering peak is at a bubble radius of about 3.2 µm, for 2 MHz at 1.5 µm. Daun concluded that "the scattering due to resonance is the dominant mechanism for scattering".

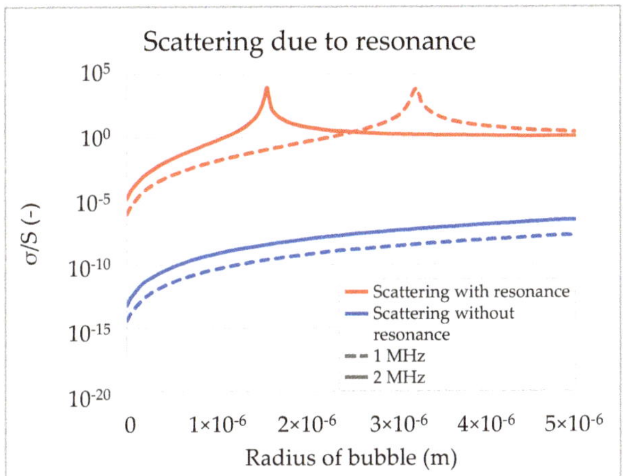

Figure 2. Scattering cross-section σ in relation to the bubble's geometric surface S as a function of the bubble radius. Resonance increases scattering effects by several magnitudes for bubbles of a critical size (adapted from [24]).

For ULP measurements this means that the signal will be affected strongly if air bubbles of a critical size are formed and entrapped during penetration. Those entrapped bubbles scatter the signal and lead to a sharp decrease in signal transmission. First, the pore size of the substrate limits the maximum possible bubble size and thus also determines if bubbles of a critical size will be created. Second, the degree of scattering always also depends on the measurement frequency as the resonance shows sharp peaks for certain bubble size–frequency combinations.

3. Fundamentals of Liquid Penetration

In order to be able to understand mechanisms that influence the ultrasound signal during penetration, a fundamental understanding of liquid penetration dynamics is necessary. In general, penetration of a liquid into a porous substrate takes place by capillary flow into the pores. Assuming cylindrical pores, the flow can be described with the Lucas–Washburn equation. It can be derived from the Poiseuille equation for laminar flow: [30]

$$\frac{dh}{dt} = \frac{r^2 \Delta p}{8\eta h},$$ (3)

where h (m) is the distance travelled by the liquid, r (m) the capillary radius, η (Ns/m²) the liquid viscosity, and Δp (N/m²) the driving pressure difference. If gravity effects are negligible—as is the case for small pores in the micrometer range like in paper [31]—the driving pressure difference is the capillary pressure. It can be described with the Laplace equation: [32]

$$\Delta p = \frac{2\gamma_L \cos\theta}{r},$$ (4)

With γ_L (N/m) being the surface tension of the liquid and θ (°) the contact angle of the liquid with the solid. Substituting Equation (4) into Equation (3) and integrating over time leads to the Lucas–Washburn Equation:

$$h = \sqrt{t} \cdot \sqrt{\frac{r\gamma_L \cos \theta}{2\eta}} \tag{5}$$

This equation indicates two important relations. First, it shows that the penetration depth (and thus the penetration volume) is proportional to the square root of time. Second, the penetration is driven by the contact angle between the liquid and the solid. Equation (4) shows that the capillary pressure is only positive if $\cos \theta > 0$, i.e., $\theta < 90°$. If $\theta > 90°$ the liquid is not sucked into the capillaries, but instead an external pressure needs to be applied so that the liquid enters the pores. Therefore, the contact angle between the liquid and the solid is the key criterion to determine if capillary penetration is possible.

Penetration into a real porous substrate is much more complex than penetration into cylindrical pores. Nevertheless, the Lucas–Washburn equation well illustrates the fundamental requirements for penetration in porous materials.

If capillary penetration is not possible, the liquid in contact with a porous substrate can still enter the pores as vapor, via surface wetting, by penetration within the porous paper fibers, or by diffusion. That is especially relevant if the substrate interacts with the liquid, as is the case for aqueous liquids and paper. The transport of water vapor through paper is most likely primarily driven by surface diffusion [30]. That means that water can enter the paper also if it is heavily sized and the contact angle is more than 90°. However, in that case there will be no liquid front filling the pores as is the case for capillary penetration. Instead water molecules will diffuse along or within the fiber surface, causing them to swell.

4. Changes in Ultrasound Transmission During Liquid Penetration

Based on the theoretical considerations made above, three cases of liquid penetration can be distinguished, each yielding a characteristic ULP curve shape. First, we distinguished between signal increase and decrease. As indicated earlier, a signal increase would be the general case to be expected if the air in the sample is displaced by the test liquid. A decrease can occur in certain cases due to entrapped air bubbles or swelling of the fibers.

4.1. Case 1—Signal Increase

A signal increase can be explained by a decrease of the reflection coefficient if the air in the sample is homogeneously replaced by the test liquid. The ultrasound signal is reflected to a large extent at air-solid boundaries and air-liquid boundaries, but much less at liquid-solid boundaries. When the sample is being filled with liquid, less and less of the signal is reflected and thus, the signal increases.

Figure 3 illustrates a liquid penetration process yielding an increasing ultrasound signal. The exemplary ULP curve is measured with a hydrophilic polyethylene membrane in contact with water. At the start the ultrasound signal is reflected to a large extent due to an air film present at the surface of the substrate (Figure 3b—image 1). When the sample is wetted by the liquid, the air film at the surface disappears and thus less reflection occurs (Figure 3b—image 2). Once the sample is soaked by the liquid, less and less reflection occurs, and the signal continues to increase. Since the sample backside is closed, the air within the sample possibly cannot escape as it is pushed back by the penetrating liquid front. Air bubbles may form. However, if the bubbles are not in a size range critical for resonance, they will not stop the signal increase. For non-resonating air bubbles to lead to a signal decrease, more air-liquid boundaries would have to form than air-solid boundaries were present at the beginning. That is not to be expected.

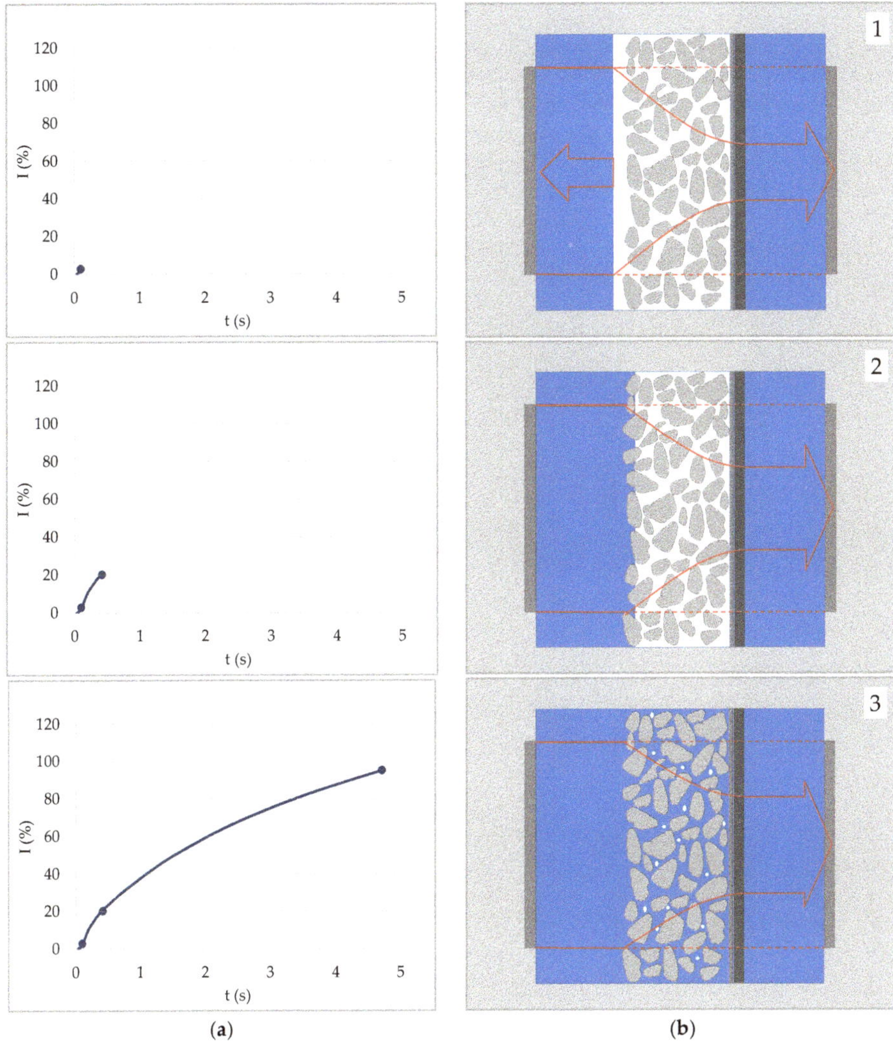

Figure 3. General case of a liquid penetrating into a porous substrate. No swelling or critical air bubbles occur. (**a**) The ULP curve of a hydrophilic polyethylene membrane in contact with water. (**b**) Graphical illustration of three stages during liquid penetration of the substrate as a function of time. Air film between liquid and surface (1), surface wetting of substrate (2), and full capillary penetration of the substrate (3).

As mentioned earlier, cases with and without capillary penetration should be distinguished. A signal increase can occur in both cases. The starting situation for both cases is the same. However, if capillary penetration is not possible, the liquid will not penetrate the substrate and thus not advance any further than in Figure 3b—image 2. Since the air film at the surface of the sample is removed nonetheless, the signal will still increase at the start also without capillary penetration. An example of such a case is a $CaCO_3$-polyethylene composite in contact with water (see Section 5).

4.2. Case 2—Signal Decrease Due to Critical Air Bubbles

In general, one would expect the ultrasound signal to increase during penetration. However, some substrates exhibited a steep decrease of the ultrasound signal immediately after getting into contact with the liquid. This is often the case for substrates where capillary penetration is possible and takes place rapidly. Such a behavior can be explained by air bubbles of a critical size. As mentioned before, if the air within the sample cannot (completely) escape, air bubbles will form during penetration. If the size of the air bubbles is in a critical range for resonance, they will drastically reduce ultrasound transmission (compare Figure 2) and cause the sharp decrease in transmission.

This behavior is typical for hydrophilic (unsized) paper in contact with aqueous liquids. Figure 4 illustrates what happens during capillary penetration of such a substrate. Wetting of the sample was so fast that it could not be detected by the ULP measurements. Penetration took place rapidly and air bubbles resonating with the signal form immediately, leading to a steep decrease of the ultrasound signal.

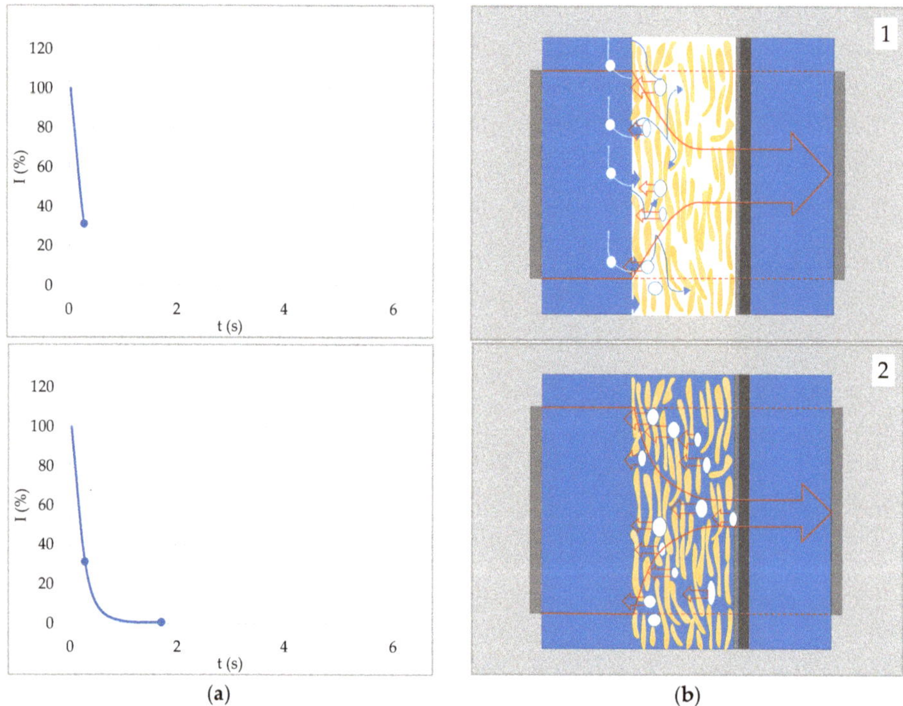

Figure 4. Capillary penetration of a liquid into a substrate (paper) with critical air bubbles being formed. (**a**) ULP curve of an unsized paper in contact with water. (**b**) Graphical illustration of two points in time during liquid penetration of the unsized paper: immediate wetting of the surface (1), air bubbles trapped in the substrate lead to ultrasound scattering, and a fast decrease of the signal (2).

In order to prove that the sharp decrease of transmitted signal is indeed caused by enclosed air bubbles, we compared measurements of hydrophilic paper with water, performed with two different sample holders. The ULP measurements discussed so far were all performed with a flat sample holder on which the samples were attached via two-sided adhesive tape (compare Figure 1c left). In this case the air cannot escape at the backside of the sample and was trapped within the substrate, where air bubbles started to form. A sample holder with grooves allowed the air to escape at the backside of the sample (compare Figure 1c right). The adhesive tape was pushed into the grooves with a grooved cylinder. Thus, the sample was in contact with the tape only at the peaks between the grooves. Still, the sample was in contact with the liquid only on the top side. When the liquid then penetrated the sample, the air was again pushed to the back, but could now escape via the grooves.

Figure 5 compares ULP measurements of unsized paper on the flat and grooved sample holder. On the flat sample holder (Figure 5a) the air could not leave the sample and air bubbles were entrapped while the air could leave on the grooved sample holder (Figure 5b). The shape of the two curves was remarkably different. While the signal decreased strongly due to the air bubbles formed in the case of the flat sample holder, there was a steep increase of transmission when the air was allowed to escape. This means that without entrapped resonating air bubbles, the signal increased due to the air being replaced by the liquid, as predicted in case 1. For the case of capillary penetration together with resonating air bubbles on a flat sample holder, on the other hand, the changes in signal transmission were controlled by the rate of formation of entrapped bubbles.

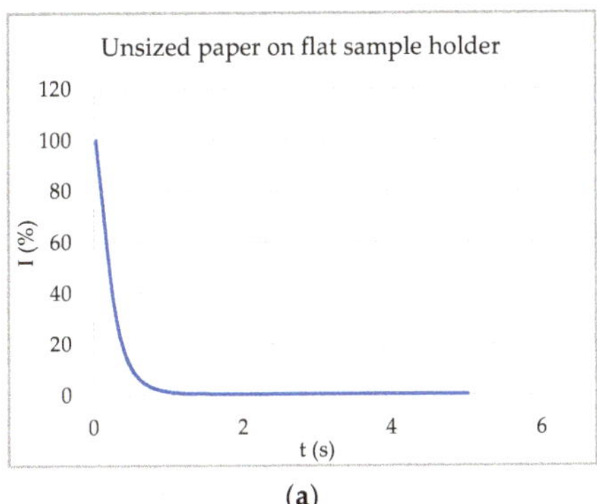
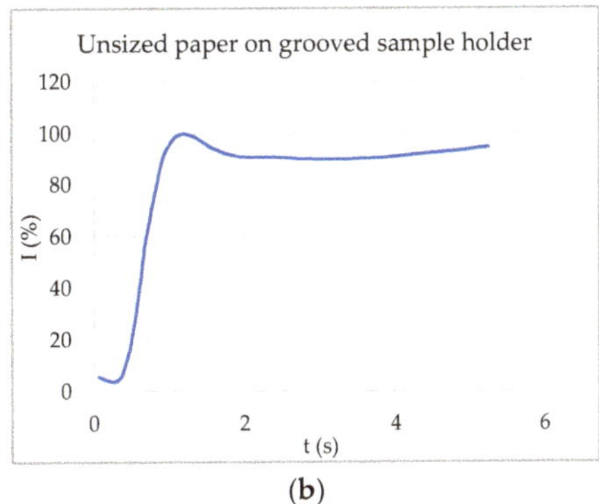

Figure 5. ULP curves of unsized paper in contact with water. (**a**) Unsized paper on a flat sample holder—air bubbles are entrapped leading to a sharp decrease of the signal. (**b**) Unsized paper on grooved sample holder (measurement from [16])—air can leave the sample on the backside via the grooves. Thus, no air bubbles form and the signal increases due to filling of the substrate pores.

4.3. Case 3—Signal Decrease Due to Swelling

If capillary penetration is possible, a signal decrease is caused by entrapped, resonating air bubbles, as explained above. However, paper samples where capillary penetration is not possible because they are heavily sized ($\theta > 90°$) can still exhibit a decrease in signal transmission. Air bubbles cannot be the reason for the decrease in this case, as there is no liquid entering the substrate pores. This can be shown when again comparing the flat and the grooved sample holder. Figure 6a shows the ULP curve of a sized copy paper with water on the flat sample holder, where the air within the sample is entrapped at the backside. The signal first increased until it reached a maximum that was followed by a slow decrease. The shape of the curve was not changed when using a grooved sample holder that allows the air to escape (Figure 6b). Thus, enclosed air bubbles could not be the reason for the signal decrease in this case.

Air bubbles cannot be the reason for a signal decrease if capillary penetration is not possible, since there is no liquid front advancing through the sample. However, as mentioned earlier, water can still enter the sample via vapor and surface diffusion, as well as penetration within the paper fibers (which are porous). In that way, water will diffuse into the fibers, filling voids within the fibers, causing them to swell. The filling of the voids should lead to an increase in the speed of sound in the fibers. That would lead to an increase in the fibers' wave impedance and consequently to an increased reflection coefficient for air–fiber boundaries. As the fibers are still surrounded by air, this would explain the slowly decreasing ultrasound signal.

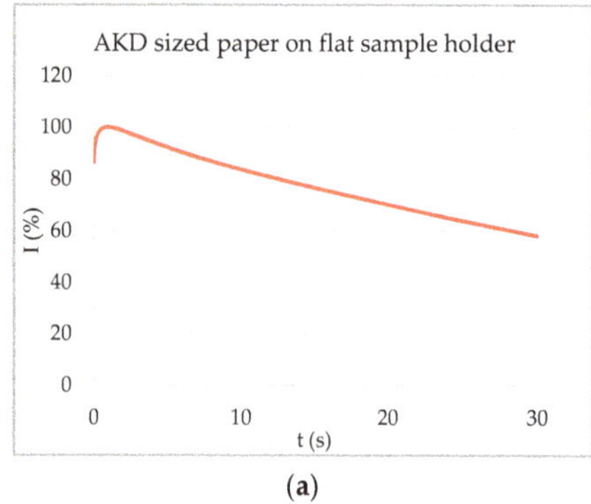
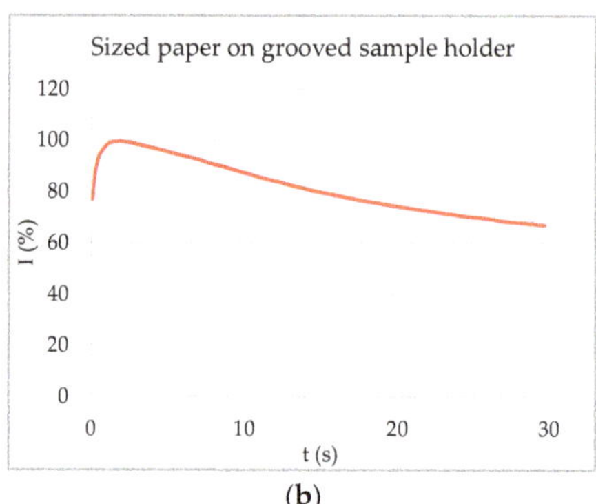

Figure 6. ULP curves of sized paper in contact with water. (**a**) AKD (alkyl ketene dimer) sized paper on flat sample holder—air is entrapped. (**b**) Sized copy paper on grooved sample holder (measurement from [16])—air can leave the sample on the backside via the grooves. Both papers have contact angles of about 100° with water, thus no capillary penetration. The signal decrease is created by fiber swelling, diffusion, or surface transport of liquid into the substrate (in this case usually paper).

Figure 7 shows the last stage of penetration of water into a sized paper. The first two stages were equivalent to the first two stages of case 1 (Figure 3—top and middle row). Initially there was an air film at the paper surface at which the ultrasound signal was reflected to a large extent. As the liquid wetted the surface and the air film disappeared, ultrasound transmission increased. For a non-swelling substrate–liquid combination the signal would not decrease again. For paper in contact with a fiber swelling liquid like water, however, the fibers will take up some liquid, ultimately leading to a decrease in the signal. In that case, the time of the maximum intensity can be interpreted as the time at which swelling of the fibers starts [33].

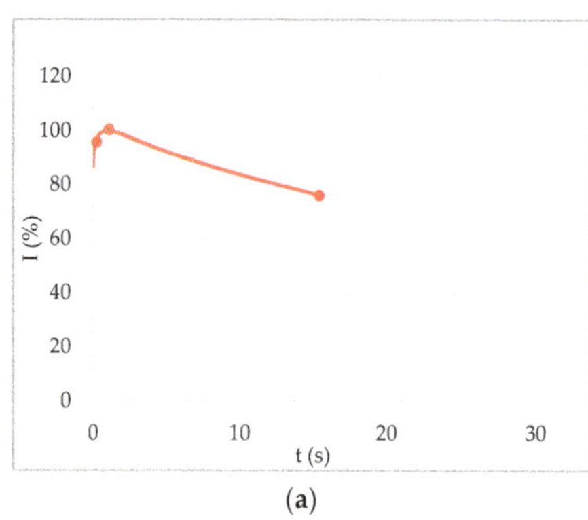

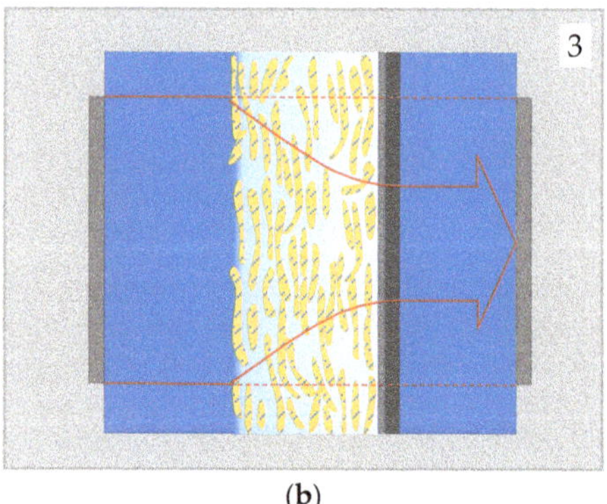

Figure 7. Liquid uptake of a surface hydrophobized paper (no capillary penetration). (**a**) ULP curve of an AKD sized paper in contact with water. (**b**) Graphical illustration of the water uptake into the fibers by vapor and surface diffusion as well as intra-fiber liquid penetration (third stage of liquid uptake; stages one and two are equivalent to Figure 3b, digits 1 and 2). The density increase within the fibers leads to a higher ultrasonic reflection at the fiber-pore interfaces and thus to a decrease of the signal.

Results that confirmed that fiber swelling was the reason for the decrease in ultrasound transmission for surface hydrophobized (sized) papers are shown in Figure 8. First, Gabriel [6] found that ultrasound transmission and wet expansion are strongly correlated especially for sized and coated papers (Figure 8a) and that they show exactly the same development over time. The wet expansion of a paper sheet is directly related to the swelling of the fibers within. Therefore, this correlation is a strong indication that the decrease of the ULP curve for hydrophobized papers is indeed related to swelling of the fibers.

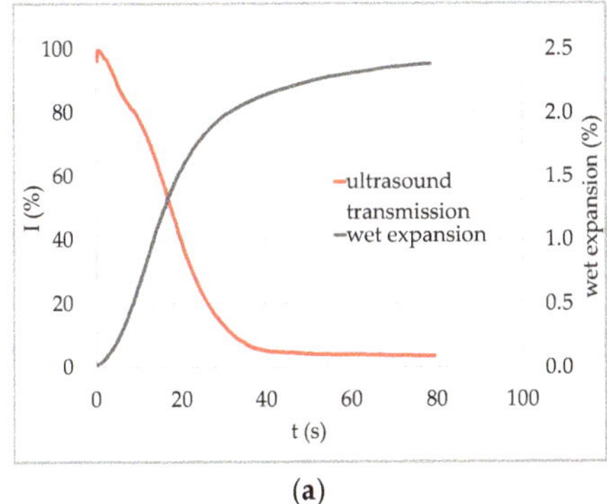

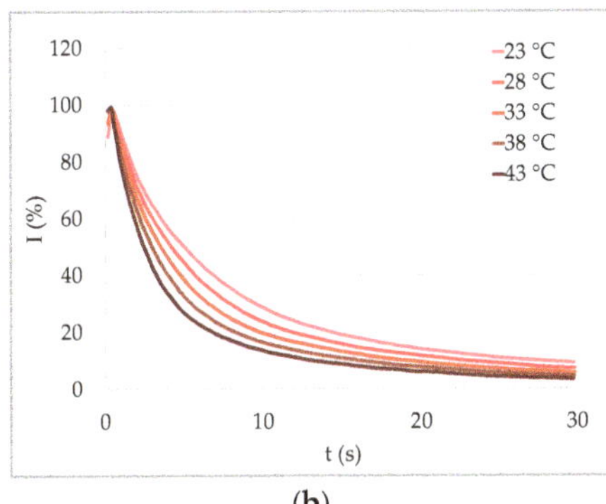

(a)   (b)

Figure 8. ULP measurement results underlining the effect of fiber swelling on ultrasound transmission. (a) Ultrasound transmission intensity I and wet expansion for a coated printing paper in contact with water. Ultrasound transmission and wet expansion are strongly correlated. (b) Temperature dependence of ULP measurements of sized copy paper in contact with water. The high temperature dependence indicates a high influence of the vapor pressure and thus diffusive water transport into the fibers (both adapted from [6]).

Second, Gabriel [6] also found that the ULP results are highly dependent on the temperature of the test liquid. In Figure 8b 20 percent ultrasound intensity was reached after about 14 s when measuring water at 23 °C, while it took less than half the time to reach the same relative intensity at 43 °C. The faster decrease at higher temperatures indicates faster water uptake of the fibers. One reason for that might be the lower viscosity of the liquid. More dramatic, however, is the increase in saturation vapor pressure, which increased exponentially with temperature. Therefore, Gabriel [6] concluded that the sharper decrease and thus faster liquid uptake at higher temperatures is caused by the intensified water vapor diffusion into the sample and the fibers. This is another indication that the decrease of the ULP curve of a surface hydrophobized (sized) paper is determined mainly by the transport of the liquid into the fibers.

5. Impact of Penetration Mechanism on Curve Shape

As discussed above, ultrasound transmission is influenced in different ways depending on whether capillary penetration of the liquid in the substrate is possible or, if not, by swelling of the substrate (often paper fibers). Each of the three discussed cases leads to a characteristic curve shape. In order to prove that capillary penetration and fiber swelling are determining the shape of the ULP curves, measurements with different test liquids were performed on two substrates. The liquids were designed so that some of them would penetrate via capillary penetration and some not. One of the substrates contains fibers that swell when in contact with aqueous liquids, the other not.

5.1. Materials and Methods

Table 3 gives an overview of the test liquids used. They are all based on deionized water with additives to modify the penetration behavior. Additives used were glycerol (AnalaR® NORMAPUR® from VWR, 99.5%), hexanediol (1,2-Hexandiol, Alfa Aesar from Thermo Fisher Scientific, 97%), and isopropanol (EMSURE®ACS from Merck, 99.8%). Glycerol was used to change the contact angle of the liquid, hexanediol and isopropanol also lowered the surface tension, which also affects the contact angle. Naphthol Blue Black (98%) was used as a dye for better visibility, but does otherwise not change the liquid properties (compare [34]). In that way, the liquids were tuned so that some of them penetrate via capillary penetration and others not. To determine whether capillary penetration is possible for a liquid–substrate combination, contact angle measurements were performed with a Fibro DAT 1100 dynamic contact angle instrument from FIBRO System, Netherlands, according to ANSI/Tappi standard T 558 om-15. Twenty individual drops with a volume of 4 µL were measured for each liquid–substrate combination. The contact angle was measured as soon as the drop had completely settled at the surface (20 ms or 40 ms after first drop-surface contact). The non-swelling substrate is a porous composite consisting of 80 percent calcium carbonate and 20 percent polyethylene. The swelling substrate is an AKD sized standard all-purpose office paper, consisting of bleached eucalyptus pulp and 14 percent calcium carbonate as filler. ULP measurements were performed with a Penetration Dynamics Analyser 2.0 from emtec Electronic GmbH, Leipzig, Germany, using a measurement frequency of 2 MHz and a flat sample holder. The samples were cut to rectangles of 7 cm × 5 cm and attached to the sample holder with two-sided adhesive tape. The ultrasonic sensor area used is a circle with a diameter of 35 mm. Each curve presented has been averaged from four individual measurement curves. Measurement time was set to 5 s.

Table 3. Test liquids and their components.

Test Liquid	Water (w%)	Glycerol (w%)	Hexanediol (w%)	Isopropanol (w%)	Dye (w%)
TL1	41.9	48	10	-	0.1
TL2	64.9	25	10	-	0.1
TL3	42.4	57.5	-	-	0.1
TL4	74.9	25	-	-	0.1
TLC	56.4	42	1.5	-	0.1
H2O	100	-	-	-	-
IPA	84	-	-	16	-

5.2. Results and Discussion

The $CaCO_3$-polyethylene composite will not swell when in contact with any of the test liquids. Therefore, in theory, only the first two cases discussed before can affect the ultrasound signal. That means that in general the signal should increase during liquid contact (case 1). A decrease is only possible if the sample is penetrated via capillary penetration and air bubbles of a critical size for resonance form (case 2).

Figure 9b shows that the test liquids 3 and 4 as well as deionized water had contact angles of more than 90° on the $CaCO_3$ -polyethylene composite sheets. That means that capillary penetration was not taking place for those liquids. The ultrasound transmission curves (Figure 9a) for these three liquids had a similar shape and all increase continuously throughout the measurement (i.e., case 1). This is what would be expected of liquids without capillary penetration on a non-swelling substrate.

The center point test liquid (TLC) and the isopropanol–water mixture (IPA) had contact angles of less than 90° and thus, capillary penetration was possible. In contrast to the other liquids, ultrasound intensity decreased for those two liquids. This indicates that air bubbles formed during the penetration of the substrate, which resonate with the signal and attenuate it. Furthermore, the TLC liquid had a higher contact angle than the IPA liquid. This is in line with the ultrasound transmission decreasing more slowly for the TLC than the IPA liquid. For a higher contact angle, the capillary pressure was lower and thus also the driving force for penetration was reduced (compare Section 3), ultimately leading to slower attenuation of the ultrasound signal.

From the measurements on the $CaCO_3$ polyethylene composite sheets it could be concluded that the shape of the ultrasound transmission curve is an indicator for the penetration mechanism taking place for non-swelling substrates, if bubbles of a critical size are formed during penetration. In the case of capillary penetration with critical bubble formation, the slope of the curve is an indicator for penetration speed.

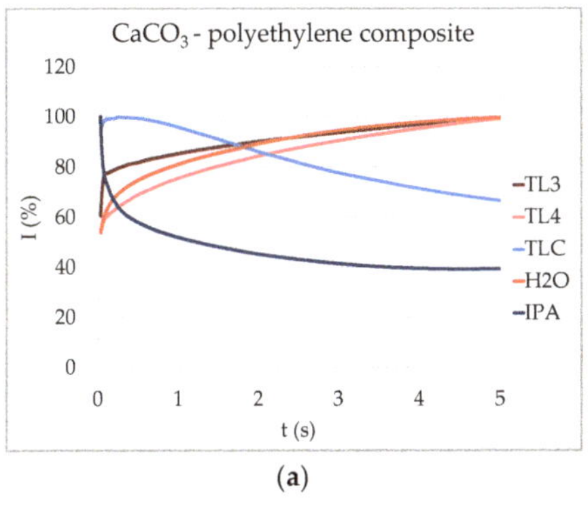
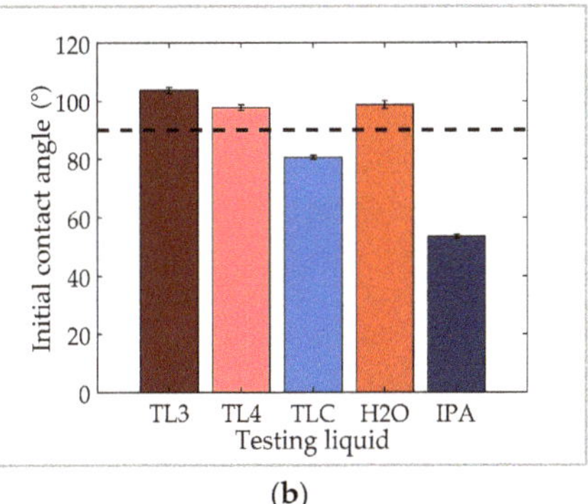

(**a**)   (**b**)

Figure 9. ULP measurements on CaCO$_3$ polyethylene composite sheets. (**a**) ULP curves; (**b**) initial contact angles θ (after 40 ms) of the testing liquids on the CaCO$_3$ polyethylene composite sheets. Liquids with capillary penetration (θ < 90°) exhibit a decreasing curve (case 2). Liquids without capillary penetration (θ > 90°) show a monotonically increasing curve; the curve is not decreasing because the substrate cannot swell (case 1; error bars indicate 95 percent confidence intervals).

As discussed before, the ultrasound signal can be affected also by changes in the fibers due to liquid uptake when it comes to paper in combination with fiber swelling liquids (case 3). This effect is especially relevant for sized papers, where the fiber surface is chemically hydrophobized, which hinders capillary penetration. Figure 10b shows that again test liquids 3 and 4, as well as water formed contact angles of more than 90° on the AKD sized paper. The TLC liquid was at the very edge of the 90° line. The shape of the ultrasound transmission curves (Figure 10a) of these four liquids matched the curve described as case 3. The increase of the ultrasound transmission at the beginning was followed by a decrease caused by diffusive liquid uptake of the fibers. For TL3 and TL4 the decrease was not very evident because it was rather slow. It would be more prominent at longer measurement intervals.

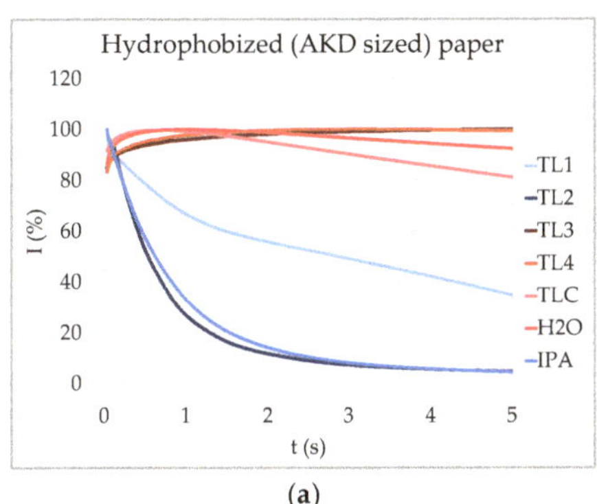
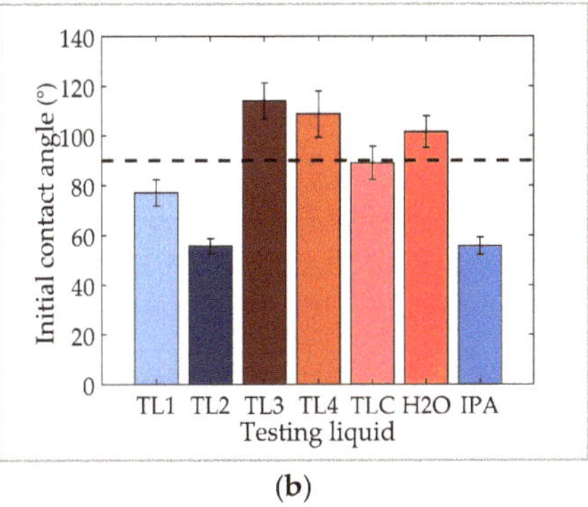

(**a**)   (**b**)

Figure 10. ULP measurements on hydrophobized (AKD sized) paper. (**a**) ULP curves and (**b**) initial contact angles θ (after 20 ms) of the testing liquids on the paper. Liquids with capillary penetration (θ < 90°) exhibit a fast decreasing curve (case 2). Those without capillary penetration (θ > 90°) show a slow decrease after a maximum, which indicates liquid uptake into the fibers due to diffusion and intra-fiber liquid transport (error bars indicate 95 percent confidence intervals.).

TL3 exhibited the highest contact angle and was also the last to reach its maximum in the ultrasound transmission. The lower the contact angle of those four liquids, the better is the surface wetting of the liquid and the faster the maximum of the curve is reached. TLC, which was at the very border of the 90° limit, was the first to reach its maximum in ultrasound transmission and the signal decreased faster after the maximum than for the other three liquids with θ > 90°. This indicates a faster liquid uptake of the fibers.

The other three test liquids—TL1, TL2, and IPA—had contact angles below 90°. For these liquids capillary penetration took place and they exhibited a curve shape, which matched case 2—capillary penetration with critical bubble formation. TL2 and IPA had a much lower contact angle than TL1 and demonstrated a steeper decrease in ultrasound transmission as well. Just like for the liquids with capillary penetration on the $CaCO_3$ polyethylene composite sheets, the slope of the curve is an indicator for penetration speed also for paper if capillary penetration takes place.

These measurements suggest that the shape of the ULP curve can be an indicator for the dominating liquid uptake mechanism for the paper. However, swelling of the fibers could also have a (minor) impact if capillary penetration is possible. There might be a transitional area close to contact angles of 90° in which both resonating air bubbles and swelling fibers affect ultrasound transmission equally. Furthermore, if paper properties vary significantly throughout the thickness direction (like is the case for highly calendared or coated papers) interpretation of ULP signals becomes even more complex. In order to avoid incorrect interpretations of ULP measurements we recommend to always perform complementary contact angle measurements.

6. Parameters for Evaluation of ULP Measurements

After elucidating the interpretation of the ULP measurement curves, it is useful to define parameters that describe specific aspects of the liquid–substrate interaction. In literature a huge variety of parameters is used. Most of the parameters used refer to curve shapes of case 3—curves with a maximum after which the signal declines. Some parameters describe the first part of the curve, which are said to describe surface wetting and surface hydrophobization:

- L = intensity difference between the initial and the maximum signal [15,16];
- G1 = slope of the curve before the maximum [16];
- W = area above the curve until the maximum [4,23].

The parameter used most often for this type of curve is the time until the maximum intensity is reached (t_{max}) [3–5,8,11–16,23,35], which often has been interpreted as the wetting time. As we pointed out, it is rather related to the onset of fiber swelling. To describe the curve after the maximum, several parameters related to the slope of the curve were used, which were interpreted as penetration speed and thus to be related to (bulk) hydrophobicity and porosity:

- A30/A60 = area above the curve after maximum until 30/60 s measurement time [3,4,23];
- Slope after maximum [8,11,12];
- T95/T60 = time until ultrasound intensity declined to 95/60 percent (after maximum) [5,13,21,22].

For the curves of case 3, we proposed to use the time until the maximum intensity to describe the first part of the curve as a measure for the time after which fibers begin to swell. The slope of the curve or a time until a certain intensity was reached (e.g., T95) should be used for the second part of the curve (Figure 11a). For a hydrophobic paper, the slope as well as T95/T60 describe the speed of water uptake of the fibers via vapor and surface diffusion.

When capillary penetration takes place and the resulting curves do not show a maximum (case 2), we recommend to use the same parameters to describe the slope of the curve as for the decreasing branch of curves with maximum (Figure 11b), i.e., T95/T60 or the slope of the curve. These parameters are a measure for the penetration speed in this case. When calculating the slope of the curve, it is important to choose a useful time interval for which the slope is calculated. Figure 11b illustrates a poorly chosen time interval. It is so large that in that interval the slopes of the two blue curves are equivalent although the dark blue curve is decreasing faster. Therefore, the time until a certain ultrasound intensity is reached (T95/T60) seems to be the more reliable parameter.

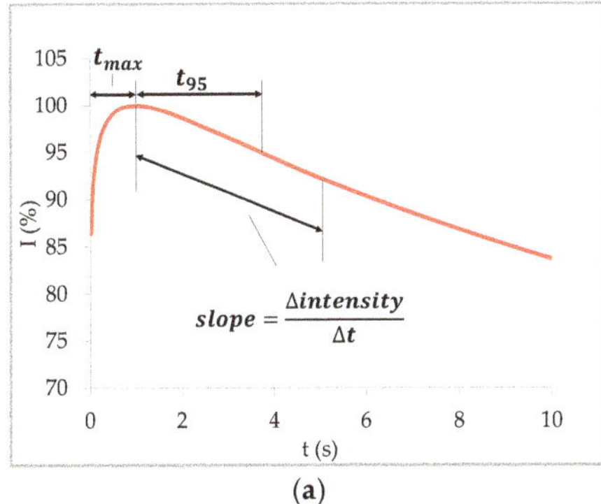

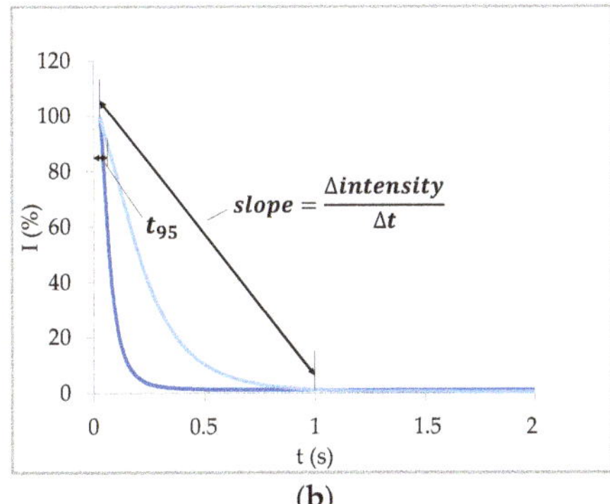

Figure 11. Proposed parameters for the evaluation of ULP measurements. (**a**) Curves with a maximum (case 3): t_{max} = time after which fiber swelling starts; slope/T95 = speed of water uptake of the fibers and (**b**) curves without maximum (case 2): slope/T95 = penetration speed.

7. Recommendations

In addition to the parameters to be used, there are some general recommendations when performing ULP measurements. First, as mentioned before, we proposed to perform contact angle measurements with the same liquids and substrates in order to find out whether capillary penetration is taking place for a liquid substrate combination. Measuring the contact angle will support interpretation of the ULP results.

Second, we wanted to point out the necessity of understanding the processes taking place during penetration in order to enable a sound interpretation of the results. If new combinations of liquids and substrates are used, it might not be clear which physical/chemical processes could influence ultrasound transmission during penetration. There might be cases in which the slope of a decreasing ultrasound curve cannot be simply interpreted as penetration speed. An example for such a case would be measurements with coagulating inks on primed surfaces, where the ink forms a coagulating layer once it gets into contact with the paper surface.

If ULP measurements of new combinations of liquids and substrates are to be performed, a comparison of measurements with a flat and a grooved sample holder might be useful. Comparing the two sample holders can reveal how strongly ultrasound transmission is influenced by entrapped air bubbles, which is an important information for the interpretation of results.

Furthermore, mostly 1 and 2 MHz have been used to perform ULP measurements and these frequencies have proved to be suitable for paper. However, other measurement frequencies could potentially reveal events on different spatial scales [10] and might be needed for substrates other than paper (e.g., membranes, nanoporous films, etc.). In this context it is important to remember that the bubble size critical for resonance also depends on the frequency of the ultrasound signal.

8. Conclusions

The main mechanisms influencing ultrasound transmission during liquid penetration were examined in order to deliver a basis for sound interpretation of ULP measurement curves. Three cases could be distinguished each yielding a characteristic curve shape:

1. In general, air being replaced by a liquid in the sample would lead to a continuous increase in ultrasound signal due to a reduction of ultrasound reflection at air–substrate boundaries.
2. If capillary penetration of the liquid into the substrate was taking place (indicated by a contact angle lower than 90°), a sharp decrease of ultrasound transmission was often observed. This was caused by enclosed air bubbles forming during penetration. If their size was in a range critical for resonance effects, the ultrasound signal was heavily scattered and thus attenuated. Therefore,

the ultrasound transmission curve in this case reflected the rate of air bubble formation during penetration, which was directly related to the penetration speed of the liquid into the paper.

3. If capillary penetration could not take place, air bubbles could not form, and the signal normally increased continuously as the air layer between liquid and substrate disappeared. For swelling substrates like paper, however, another mechanism was taking place. The typical measurement curve exhibited an increase up to a maximum followed by a decrease in ultrasound transmission. The signal first increased due to liquid replacing the air layer at the paper surface. After some time, the substrate (fibers) started to take up water via vapor and surface diffusion, which caused a density increase and thus higher reflection at the air–fiber interfaces, which resulted in a decreasing signal. The decrease in ultrasound transmission was proportional to water uptake of the fibers in this case. This means that the time until the maximum was related to the onset of fiber swelling. A shorter time to maximum and faster decrease after the maximum means faster swelling.

With testing liquids tailored to show (or not show) capillary penetration, as indicated by the contact angle, we were able to prove that a fast falling ULP curve without a maximum indeed indicates the presence of capillary penetration in the substrate. By comparing the curves between non-swelling and swelling substrates in the absence of capillary penetration we could demonstrate that the water uptake in the substrate material (in this case the liquid uptake into the paper fibers due to diffusion and intra-fiber wetting) corresponds to the slope of the signal decrease after the maximum in the curve. These results suggest that ULP measurements together with contact angle measurements could be a means to quantify first capillary penetration into substrate pores, and second diffusion induced liquid uptake in the substrate material (in the absence of capillary penetration).

Author Contributions: Conceptualization, C.W. and U.H.; investigation, C.W.; data curation, C.W.; writing—original draft preparation, C.W. and U.H.; writing—review and editing, C.W. and U.H.; visualization, C.W.; supervision, U.H.; project administration, U.H.; funding acquisition, U.H. All authors have read and agreed to the published version of the manuscript.

Funding: The financial support by the Austrian Federal Ministry for Digital and Economic Affairs and the National Foundation for Research Technology and Development is gratefully acknowledged. We also thank our industrial partners Mondi, Canon Production Printing, Kelheim Fibres, and SIG Combibloc for their financial support.

Acknowledgments: The authors wish to thank Sarah Krainer for interesting discussions and contact angle measurements.

References and Note

1. Pan, Y.; Kuga, S.; Usuda, M.; Kadoya, T. Ultrasound can evaluate paper sizing. *Tappi J.* **1985**, *68*, 98–102.
2. Pan, Y.; Kuga, S.; Usuda, M. An ultrasonic technique to study wetting and liquid penetration of paper. *Tappi J.* **1988**, *72*, 119–123.
3. Hutton, B.; Shen, W. Sizing effects via AKD vaporisation. *Appita J.* **2005**, *58*, 367–373.
4. Grenz, R. *Reduzierung der Additivkosten bei der Herstellung von massgeleimten Papieren durch Identifizieren, Quantifizieren und Minimieren der negativen Wechselwirkungen von Leimungsmitteln mit Retentionsmitteln und optischen Aufhellern*; Papiertechnische Stiftung: Heidenau, Germany, 2009; pp. 1–60.
5. Martorana, E.; Fischer, S.; Kleemann, S. Quantitative analysis of synthetic sizing agents (ASA/AKD) using NIR spectroscopy. *Nord. Pulp Pap. Res. J.* **2009**, *24*, 335–341. [CrossRef]
6. Gabriel, G. Messtechnische Erfassung der Wechselwirkungen zwischen Flüssigkeiten und Papieren Mittels Ultraschall. Ph.D. Thesis, Technische Universität Graz, Graz, Austria, 1999.
7. Hernádi, A.; Lele, I.; Rab, A. Correlation between initial wettability and sheet making properties of secondary fibres. *Papiripar* **2003**, *47*, 167–171.
8. Sarah, K.; Ulrich, H. Short timescale wetting and penetration on porous sheets measured with ultrasound, direct absorption and contact angle. *RSC Adv.* **2018**, *8*, 12861–12869. [CrossRef]

9. Przybysz, K.; Drzewińska, E.; Stanisławska, A.; Wysocka-Robak, A.; Cieniecka-Rosłonkiewicz, A.; Foksowicz-Flaczyk, J.; Pernak, J. Ionic liquids and paper. *Ind. Eng. Chem. Res.* **2005**, *44*, 4599–4604. [CrossRef]

10. Stor-Pellinen, J.; Hæggström, E.; Luukkala, M. Measurement of paper-wetting processes by ultrasound transmission. *Meas. Sci. Technol.* **2000**, *11*, 406–411. [CrossRef]

11. Bordenave, N.; Grelier, S.; Coma, V. Hydrophobization and antimicrobial activity of chitosan and paper-based packaging material. *Biomacromolecules* **2010**, *11*, 88–96. [CrossRef]

12. Bordenave, N.; Grelier, S.; Pichavant, F.; Coma, V. Water and moisture susceptibility of chitosan and paper-based materials: Structure-property relationships. *J. Agric. Food Chem.* **2007**, *55*, 9479–9488. [CrossRef]

13. Wan, Z.; Wang, L.; Yang, X.; Guo, J.; Yin, S. Enhanced water resistance properties of bacterial cellulose multilayer films by incorporating interlayers of electrospun zein fibers. *Food Hydrocoll.* **2016**, *61*, 269–276. [CrossRef]

14. Padberg, J.; Bauer, W.; Gliese, T. Measuring the oil and grease barrier properties of MFC coated paper-using ultrasonic signal variation. *TAPPI Int. Conf. Nanotechnol. Renew. Mater.* **2017**, *2*, 702–717.

15. Oliver, J.F.; D'Souza, E.; Hayes, R.E. Application of Ultrasonic and Porometric Techniques to Measure liquid Penetration in Digital Printing Papers. In Proceedings of the IS&T's NIP18: 2002 International Conference on Digital Printing Technologies, San Diego, CA, USA, 29 September–4 October 2002.

16. Sharma, N. Use of Ultrasound to Predict Ink-Jet Print Quality. Master's Thesis, Miami University, Oxford, OH, USA, 2005.

17. Leks-Stępień, J. Effects of the coating mixture on the penetration of liquids into paper. *Fibres Text. East. Eur.* **2019**, *137*, 107–113. [CrossRef]

18. Krainer, S.; Saes, L.; Hirn, U. Predicting inkjet dot spreading and print through from liquid penetration- and picoliter contact angle measurement. *Nord. Pulp Pap. Res. J.* **2020**, *35*, 124–136. [CrossRef]

19. Dohr, C.A.; Hirn, U. Influence of paper properties on adhesive strength of starch gluing. *Packag. Sci. Technol.* **2020**. accepted.

20. Gigac, J.; Kasajová, M.; Maholányiová, M.; Stankovská, M.; Letko, M. Prediction of surface structure of coated paper and of ink setting time by infrared spectroscopy. *Nord. Pulp Pap. Res. J.* **2013**, *28*, 274–281. [CrossRef]

21. Gigac, J.; Stankovská, M.; Fišerová, M. Improvement of oil and grease resistance of cellulosic materials. *Wood Res.* **2018**, *63*, 871–886.

22. Gigac, J.; Stankovská, M.; Kasajová, M. Interactions between offset papers and liquids. *Wood Res.* **2011**, *56*, 363–370.

23. Woods, W.P. Effect of Calendaring on Flexographic Printability of Linerboard. Master's Thesis, Western Michigan University, Kalamazoo, MI, USA, 1999.

24. Daun, M. Model for the Dynamics of Liquid Penetration into Porous Structures and Its Detection with the Help of Changes in Ultrasonic Attenuation. Ph.D. Thesis, Technische Universität Darmstadt, Darmstadt, Germany, 2006.

25. Humphrey, V.F. Ultrasound and matter—Physical interactions. *Prog. Biophys. Mol. Biol.* **2007**, *93*, 195–211. [CrossRef]

26. McClements, D.J. Ultrasonic characterisation of emulsions and suspensions. *Adv. Colloid Interface Sci.* **1991**, *37*, 33–72. [CrossRef]

27. Šutilov, V.A. *Physik des Ultraschalls*; Springer: Vienna, Austria, 1984; ISBN 978-3-7091-8751-7.

28. Elsayad, K.; Urstöger, G.; Czibula, C.; Teichert, C.; Gumulec, J.; Balvan, J.; Pohlt, M.; Hirn, U. Mechanical properties of cellulose fibers measured by brillouin spectroscopy. *Cellulose* **2020**, *27*, 4209–4220. [CrossRef]

29. Karppinen, T.; Montonen, R.; Määttänen, M.; Ekman, A.; Myllys, M.; Timonen, J.; Hæggström, E. Evaluating pulp stiffness from fibre bundles by ultrasound. *Meas. Sci. Technol.* **2012**, *23*, 065603. [CrossRef]

30. Kettle, J. Chapter 7: Moisture and Fluid Transport. In *Paper Physics*; Kaarlo, N., Ed.; Finnish Paper Engineers' Association: Helsinki, Finland, 2008; pp. 265–294. ISBN 978-952-5216-29-5.

31. Schoelkopf, J.; Ridgway, C.J.; Gane, P.A.C.; Matthews, G.P.; Spielmann, D.C. Measurement and network modeling of liquid permeation into compacted mineral blocks. *J. Colloid Interface Sci.* **2000**, *227*, 119–131. [CrossRef] [PubMed]

32. Lyne, M.B. Wetting and the penetration of liquids into paper. In *Handbook of Physical Testing of Paper*; Borch, J., Lyne, M.B., Mark, R.E., Habeger, C.C., Murakami, K., Eds.; Marcel Dekker Inc.: New York, NY, USA, 2002; pp. 303–332. ISBN 0-8247-0499-1.

Materials **2020**, *13*, 2754

33. emtec Electronics Emtec Measuring Systems: New measuring methods to study the quality of paper and process liquids, their interaction and its influence to converting processes; 2002.
34. Krainer, S.; Smit, C.; Hirn, U. The effect of viscosity and surface tension on inkjet printed picoliter dots. *RSC Adv.* **2019**, *9*, 31708–31719. [CrossRef]
35. Yang, L.; Jianghao, L.; Lingya, G. Detailed insights to liquid absorption and liquid—Paper interaction. In Proceedings of the Transactions of the XVth Fundamental Research Symposium, Cambridge, UK, 9–13 September 2013; I'Anson, S.J., Ed.; FRC: Manchester, UK, 2018; pp. 585–598.

Article

Ultrasound Axicon: Systematic Approach to Optimize Focusing Resolution through Human Skull Bone

Fabián Acquaticci [1,2,*], **Sergio E. Lew** [1] **and Sergio N. Gwirc** [3,*]

[1] Instituto de Ingeniería Biomédica, Universidad de Buenos Aires, Buenos Aires C1063ACV, Argentina; slew@fi.uba.ar
[2] Instituto Nacional de Tecnología Industrial, Ministerio de Producción y Trabajo, San Martín, Buenos Aires B1650WAB, Argentina
[3] Departamento de Investigaciones Tecnológicas, Universidad Nacional de La Matanza, San Justo, Buenos Aires B1754JEC, Argentina
[*] Correspondence: facquaticci@inti.gob.ar (F.A.); sgwirc@unlam.edu.ar (S.N.G.)

Received: 16 August 2019; Accepted: 17 October 2019; Published: 20 October 2019

Abstract: The use of axicon lenses is useful in many high-resolution-focused ultrasound applications, such as mapping, detection, and have recently been extended to ultrasonic brain therapies. However, in order to achieve high spatial resolution with an axicon lens, it is necessary to adjust the separation, called stand-off (δ), between a conventional transducer and the lens attached to it. Comprehensive ultrasound simulations, using the open-source k-Wave toolbox, were performed for an axicon lens attached to a piezo-disc type transducer with a radius of 14 mm, and a frequency of about 0.5 MHz, that is within the range of optimal frequencies for transcranial transmission. The materials properties were measured, and the lens geometry was modelled. Hydrophone measurements were performed through a human skull phantom. We obtained an initial easygoing design model for the lens angle and optimal stand-off using relatively simple formulas. The skull is not an obstacle for focusing of ultrasound with optimized axicon lenses that achieve an identical resolution to spherical transducers, but with the advantage that the focusing distance is shortened. An adequate stand-off improves the lateral resolution of the acoustic beam by approximately 50%. The approach proposed provides an effective way of designing polydimethylsiloxane (PDMS)-based axicon lenses equipped transducers.

Keywords: ultrasonic lens; axicon lens; focused ultrasound; transcranial ultrasound

1. Introduction

Focused ultrasound (FUS) in pulsed mode is unique among transcranial brain stimulation methods in combining exceptional spatial resolution (on the millimeter scale) [1–3] with the potential to target subcortical structures (deeper than 10 cm) [4] through the intact skull. It also has potential for inducing neuronal excitation or suppression without evidence of tissue damage [5].

Recently, we demonstrated the advantages of focusing ultrasound (US) through polydimethylsiloxane (PDMS)-based axicon lenses to selectively drive brain activity [6]. The ultrasound axicon is shown in Figure 1. It has the shape of a cone. As the cone angle (ϕ) decreases, the focus moves closer to the lens. Developing low-intensity applications include the opening of the blood–brain barrier and ultrasonic neuromodulation. Both techniques have recently been extended to human subjects and are under active research. Spherically FUS transducer has been, until now, the most commonly used for transcranial focusing ultrasound, but it has a large focal length (several centimeters) that may hinder its coupling with the head, for example, when the focal plane is too close to skull inner ivory layer. With spherical segment ultrasound transducers, FUS is commonly delivered through a big plastic bag containing degassed water placed over the scalp. This is due to the fact that, for cortical stimulation, the acoustic beam should be focused a few millimeters deep from the skull surface [3]. In this sense, the great advantage of the axicon lens is its ability to suppress the near-field and maintain a very near focus from the lens face outward. PDMS-based axicon lens affixed on the face of conventional transducers makes it possible to build more compact devices without liquids that can regasify or leak, with a greater spatial resolution.

However, for an axicon lens to offer high spatial resolution and depth, control is necessary to adjust the separation between the transducer and axicon lens, called stand-off (δ), for each particular case [7]. Given the geometry of the lens, if δ is not properly adjusted, internal reflections can occur, making the configuration useless as a focused transducer. The lens reduces the focal length (F) from the near-field distance (N) of the attached US transducer. The relation between axicon lens angle and the value of F/N produced was described in [8] for acrylic plastic/oil combination, but the effect of the stand-off was not modeled and the adequate value of δ was not described elsewhere. In [7], the stand-off of the different settings was adjusted experimentally to obtain the best signal-to-noise ratio (SNR) for ultrasonic contact inspection. In our approach, time domain ultrasound simulations, based on the k-space pseudo-spectral methods, play a key role in enabling the modeling and systematic design of ultrasonic axicon lenses with optimum stand-off for high-resolution ultrasonic applications, as transcranial stimulation, in high-resolution mapping, and in the evaluation of a wide variety of defects. Finally, the effects of human skull on axicon fields were tested.

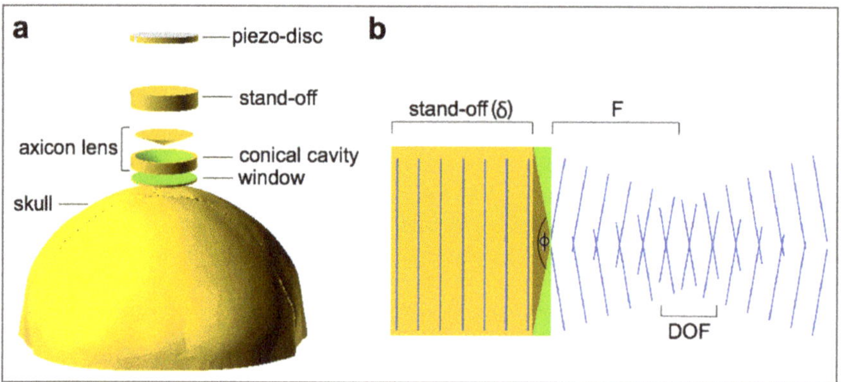

Figure 1. (**a**) Exploded model of the axicon lens. The polydimethylsiloxane (PDMS) fills the conical cavity of the lens and the stand-off. An ultrasonic lens is formed by the PDMS/plastic interfaces. The window is a thin material layer between the conical cavity and the face of the lens; (**b**) schematic ultrasonic beam pattern for transducers with axicon lenses. These are described by its total included cone angle (ϕ). Depth of focus (DOF) is the focal region and F is the desired focal length. Both DOF and F depend on the sound velocity into the material used for the interface.

Skull is generally constituted of three relatively homogeneous layers: The outer and inner tables of solid ivory bone, and the central layer of diploe of cancellous bone, with a blood- and fat-filled porous structure. The dimensions of the blood- and fat-filled inclusions are random, and an average thickness in the direction normal to the surface is about 0.6 mm. [5], so the fundamental frequency used was chosen to help alleviate the concerns for acoustic energy absorption or refraction by the skull. For frequencies lower than about 0.5 MHz, reflection of sound is the principal cause of insertion loss. At frequencies between about 0.6 and 0.9 MHz, the absorption loss in the diploe layer begins to limit sound transmission, so the oscillations are damped out and the insertion loss increases linearly with frequency. At about 0.9 MHz, the scattering loss in the diploe layer begins to limit sound transmission, so the insertion loss begins to increase as the fourth power of frequency [5].

A full-wave nonlinear ultrasound model based on the k-space pseudo-spectral method was developed and released as part of the open-source k-Wave Acoustics Toolbox [9,10]. This model can account for the propagation of nonlinear ultrasound waves in homogeneous or heterogeneous media with power acoustic absorption and without restrictions on the directionality of the waves. The accuracy of the implementation of nonlinear ultrasound model in the k-Wave Toolbox was validated using experimental measurements of the ultrasound made with a linear diagnostic ultrasound probe and a membrane hydrophone [11].

A computational model for elastic wave propagation in heterogeneous media can be constructed based on the solution of coupled first-order acoustic equations given in Equations (1)–(3) using the Fourier pseudo-spectral method. This uses the Fourier collocation spectral method to compute spatial derivatives, and a leapfrog finite-difference scheme to integrate forwards in time. Using a temporally and spatially staggered grid, the field variables are updated in a time stepping [12,13]. To simulate free-field conditions, a perfectly matched layer (PML) is also applied to absorb the waves at the edge of the computational domain [14]. Without this boundary layer, the computation of the spatial derivate via the FFT causes waves leaving one side of the domain to reappear at the opposite side. The use of the PML thus facilitates infinite domain simulations without the need to increase the size of the computational grid.

$$\frac{\partial u}{\partial t} = -\frac{1}{\rho_0}\nabla p - \alpha \cdot u \tag{1}$$

$$\frac{\partial \rho_x}{\partial t} = -\rho_0 \frac{\partial u_x}{\partial x} - \alpha_x \rho_x \tag{2}$$

$$p = c_0^2 \sum \rho_{x,y,z} \tag{3}$$

where u is the acoustic particle velocity, ρ_0 is the medium density, ρ is the acoustic density, c is the thermodynamic sound speed, p is the acoustic pressure, and $p_0 = p(t = 0)$ is the initial pressure distribution.

There are two main stages in this work: The definition of general design equations for PDMS-based axicon lenses with optimal stand-off, through comprehensive ultrasound simulations, for an optimal transcranial focusing; and the measurements and simulations performed to determine focusing performance of the proposed lenses, through a human skull phantom.

2. Materials and Methods

In order to develop a design model, axicon lenses with ϕ angles between 80° and 170° in steps of 5° were simulated with increments of δ in steps of $\lambda/4$ with 8 grid points per wavelength (λ) in the stand-off medium. The final pressure field along with the RMS beam pattern was calculated. The normalized cross-section profile area at the focus F was determined, for each δ. The minimum area represents the greatest improvement with the optimum stand-off, which minimizes transmitted energy outside of the main beam and improves lateral spatial resolution with lower possible sidelobes.

The domain was discretized using a grid point spacing of 250 μm (giving a maximum supported frequency of 2.06 MHz), and a grid size of 512 × 512 grid points (corresponding to a domain size of 128 × 128 mm). Simulations were run on an NVIDIA® GTX 950 graphics processing unit (Santa Clara, CA, United States) using the MATLAB® Parallel Computing Toolbox (Natick, MA, United States). The simulation of all angles/stand-off combinations can be completed in approximately 60 h. By default, numbers in MATLAB® are stored in double precision. However, in almost all cases, k-Wave does not require this level of precision. In particular, the performance of the PML generally limits the accuracy to around 4 or 5 decimal places. We use a PML thickness of 20 grid points that gives a transmission coefficient of −100 dB. This corresponds to a reduction in the signal level of 1×10^{-5}, which is significantly less than double precision. Further, there will also be uncertainties in the definition of the materials properties. A list of the main simulation inputs is given in Table A1. In this work, a heterogeneous medium was defined as a layered interface on both sides of the conical cavity of the axicon lens as shown in Figure A1. The convective nonlinear effects from the convection of mass was considered. However, at low frequencies and amplitudes, nonlinearity will only have a small effect on the wave field. At higher frequencies and amplitudes, this effect become more important.

The accuracy of the implementation of ultrasound model with the k-Wave Toolbox was validated in our previous work [6] using experimental measurements of FUS made with the same axicon lens attached transducer and a needle hydrophone (Force Technology MH28) within a 6-L anechoic test tank. Our previous study has already shown that there is a good agreement between the simulated and experimental beam patterns. In this work, we also characterized the acoustic pressure amplitude of the beam pattern of the axicon lens when FUS was transmitted through a human skull phantom for experimental validation of simulated transcranial ultrasound propagation. There is negligible conversion to shear waves in the layers of skull when the incidence angle is within about 20° of normal [5]. The ability of bone to support shear waves can affect transcranial transmission, although the changes to the intracranial field are typically negligible for ultrasound applied at normal or near-normal incidence. Therefore, we will model only longitudinal waves. The phantom was created from the parietal portion of a mesh segmented from MRI head image data. Clear Med610 3D printing resin was used to create the skull bone phantom. The acoustic properties of Stratasys™ materials were recently reported in [15]; thus, these measurements were not repeated as part of the current work. The reported and measured property values, and estimated uncertainty in those measurements, are shown in Table 1.

To test the effects of a human skull on FUS fields, we inserted a 5 mm thick fragment of parietal bone phantom between the transducer and the hydrophone, as shown in Figure 2. The transducer described in our previous work [6] has an ultrasonic piezo-disc-type element of 28 mm diameter (SMD28T21F1000R, Steminc Steiner & Martins, Inc., Davenport, FL, USA) of PZT-4 mounted on stainless-steel housing operating in thickness mode vibration at 445 kHz. Epoxy (Resoltech 1040, Resoltech, Rousset, France) resin was used in order to build the conical cavity of the lens with an angle ϕ of 144°. Degassed PDMS (Sylgard 184, Dow Corning, Midland, MI, USA) was used to fill the conical cavity of the lens and for the lens-transducer interface with a stand-off δ of 30 mm. The ultrasonic lens is formed by the epoxy/PDMS interface. The specifications are summarized in Table 2.

Table 1. Compressional and shear speed, attenuation, and density of Clear Med610 material.

Comp. Speed (ms^{-1})	Shear Speed (ms^{-1})	Absorption (dB cm^{-1})	Density (kg m^{-3})
2495 ± 8	1081 ± 31	3.70 ± 0.1	1180

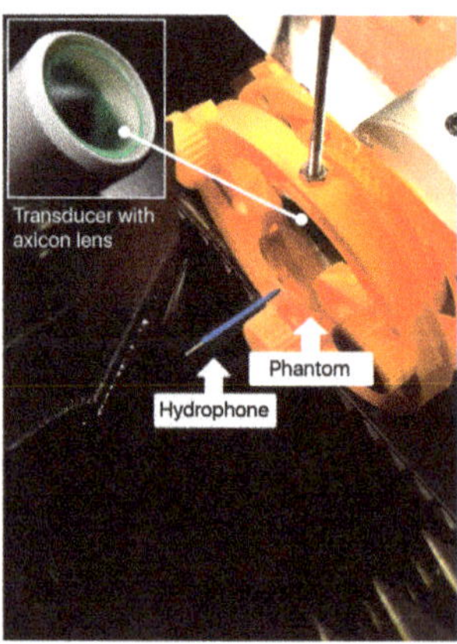

Figure 2. Photograph of the ultrasound test tank showing the axicon lens equipped transducer detail, parietal bone phantom, and hydrophone.

Table 2. Specifications of the axicon lens characterized.

Parameter	Value
Transducer frequency (f)	0.445 MHz
Transducer diameter (D)	28 mm
Transducer near field in water (N)	64.75 mm (−6 dB)
Ratio of F/N (F/N)	0.16
Axicon lens angle (ϕ)	144°
Focal length (F)	10.5 mm
Depth of focus (DOF)	11 mm (−3 dB)/22 mm (−6 dB)
Focus diameter (d_F)	2.5 mm (−3 dB)/3.5 mm (−6 dB)
Insertion loss (IL)	8.4 dB
Stand-off (δ)	30 mm
Encapsulation Echo Reduction (ER)	−12 dB

3. Results

3.1. Estimation of the Axicon Lens Angle

The relation between axicon lens angle ϕ and the value of F/N produced (see Appendix A) is illustrated graphically in Figure 3. The transducer near field length N is given by $N = D^2 f/4c$. The coefficient of determination R^2 is 0.97 with *p*-value significance level of < 0.00001. This relation is for $\delta = \delta_{Optimum}$:

$$\phi = \frac{9.9708 + \ln\left(\frac{F}{N}\right)}{5.52 \cdot 10^{-2}} \ [degrees]. \tag{4}$$

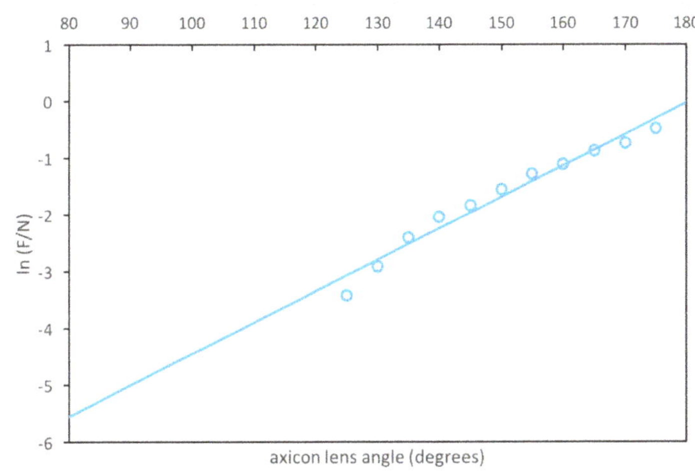

Figure 3. Illustration of the relation between the axicon lens angle ϕ and the ratio of *F/N*.

The following relation, based on our study summarized in Table A2 (see Appendix A), was found experimentally valid for the lens described. The optimum ratio of F/N appears to be between 0.1 and 0.3. This produces focal beam diameters (d_F with the axicon lens equipped transducers and d_N of a conventional transducer), and depths of focus (DOF) of similar ratio according to:

$$\frac{F}{N} = \frac{d_F}{d_N} = \frac{DOF_F}{DOF_N}. \tag{5}$$

For values of F/N > 0.4, some evidence of the original near field still remains. As the value F/N decreases below 0.4, all evidence of the original near field is rapidly suppressed in the lens system. This contrasts with the behavior of spherical lenses where some original near field is always present.

We observe that there is an inversely proportional relationship between F/N and $\kappa\delta$ values, as shown in Figure 4 for different values of angle ϕ. The relative percentage loss of lateral resolution (LLR) as a function of $\kappa\delta$ is shown in the same figure, where κ is the wave number. LLR is relative to the best lateral resolution that can be achieved with $\kappa\delta$ value that minimizes energy transmitted outside of the narrowest possible main beam. Since the speed of sound in PDMS is lower than in water, as the value of δ increases, the wavelengths-weighted average sound speed through the heterogeneous medium with step discontinuity of velocity decreases, which effectively reduces the ratio of F/N.

3.2. Reflectivity Effect on the Internal Walls of the Lens Housing

To illustrate the effect of outer case and inner isolation (Figure 5) on the lateral resolution of the lens, the relative percentage LLR as a function of $\kappa\delta$, and the normalized lateral beam profile are shown in Figure 6 for two different housing. With a reflecting housing, without inner sleeve, there will be more internal reflections in the lens system, since there are abrupt transitions of acoustic impedance with the inner walls. As a result, the lateral resolution of the lens decreases, compared to non-reflective housing with inner isolation. For example, with a PDMS-based lens at 445 kHz and a cone angle of 130°, assembled with the optimum stand-off inside a housing of Inconel-625 with internal reflectivity of 0.8 dB down, the relative LLR value is reduced by half and energy transmitted outside the main beam is 10% higher, compared to inner sleeve reflectivity of 40 dB down.

3.3. Estimation of the Optimum Stand-Off

We found a numerical relationship for δ based on the simulation results of 2420 angle/stand-off combinations for axicon lenses (see Appendix A). The relation between the optimum value of δ (which improves spatial resolution) and the value of F/N is shown in Figure 7. The linear regression equation is given by:

$$\delta_{Optimum} = \frac{10.921 + \ln\left(\frac{F}{N}\right)}{1.96 \cdot 10^2} \; [meters].\tag{6}$$

The coefficient of determination R^2 is 0.96 with a p-value significance level of < 0.00001.

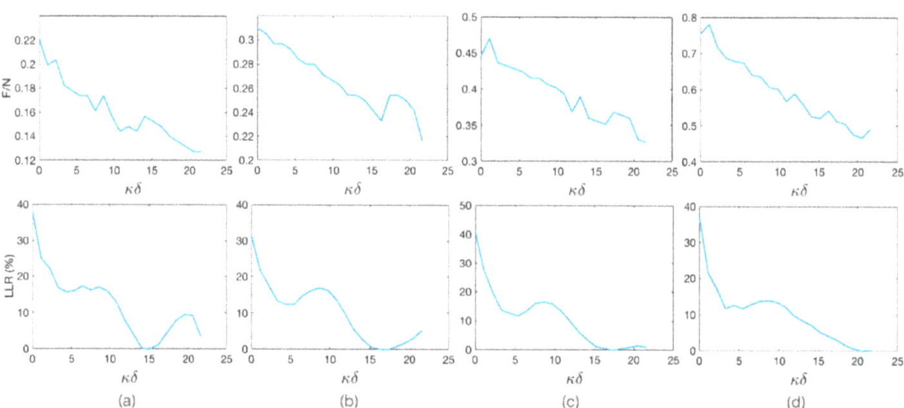

Figure 4. (Top) Value of F/N and (Bottom) relative loss of lateral resolution, both, vs. stand-off given in number of wavelengths ($\kappa\delta$) for 445 kHz PDMS-based axicon lens with different angles. (**a**) $\phi = 140°$; (**b**) $\phi = 150°$; (**c**) $\phi = 160°$; (**d**) $\phi = 170°$.

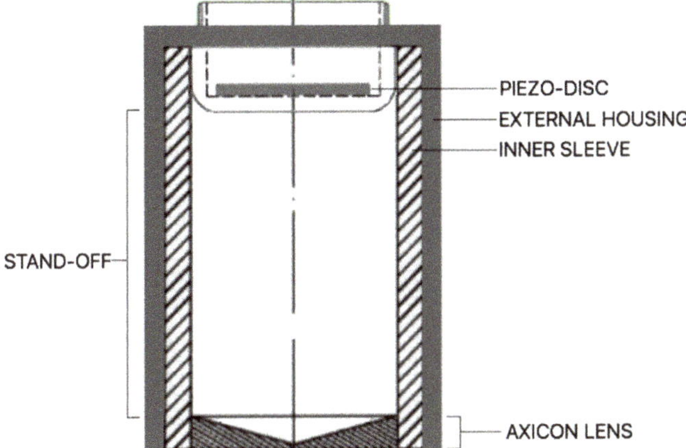

Figure 5. Housing of the axicon lens.

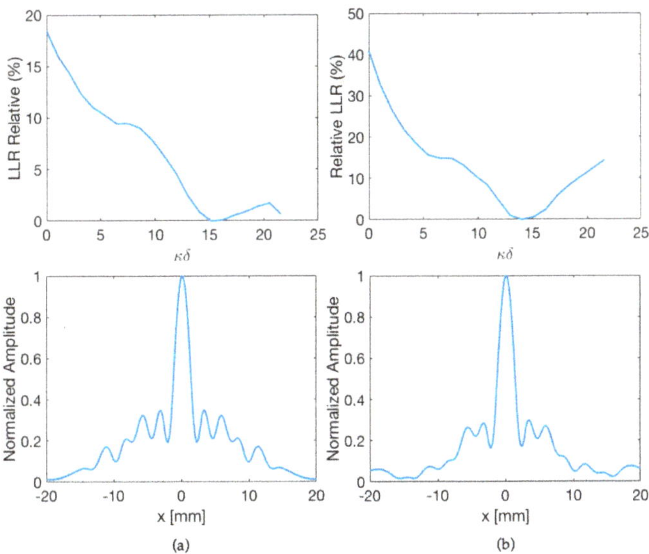

Figure 6. Outer case effect for 445 kHz PDMS-based axicon lens with cone angle $\phi = 130°$. (Top) Relative loss of lateral resolution vs. stand-off given in number of wavelengths ($\kappa\delta$) and (Bottom) normalized pressure amplitudes of the lateral beam profile. (**a**) Lens housing with reflectivity of 0.8 dB down; (**b**) lens housing with reflectivity of 40 dB down.

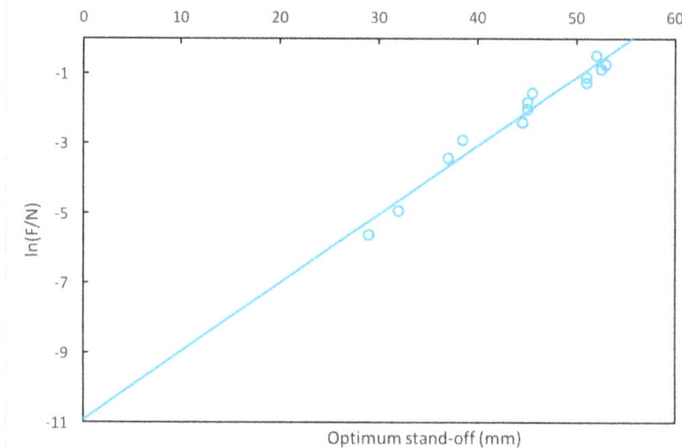

Figure 7. Illustration of the relation between the axicon lens optimum stand-off δ and the ratio of F/N. The value of δ estimated by linear regression is indicated to obtain the highest lateral resolution for different lens angles.

As an example, Figure 8 shows the focusing behavior of PDMS-based 144° axicon lens with four different values of δ (a: 0 mm, b: 20.5 mm, c: 34 mm, and d: 45.25 mm). These different settings are indicated in Figure 9 showing the relative LLR as a function of $\kappa\delta$. With a stand-off of 34 mm, the lateral spatial resolution improves by up to 40%, compared to the same lens without stand-off.

The optimum stand-off predicted by Equation (6) was checked for different frequencies. Figure 10 compares the experimental loss of lateral resolution (LLR) for transducers with acoustic frequencies (f) of (a) 0.2225 MHz, (b) 0.445 MHz, (c) 0.890 MHz, and (d) 4.45 MHz.

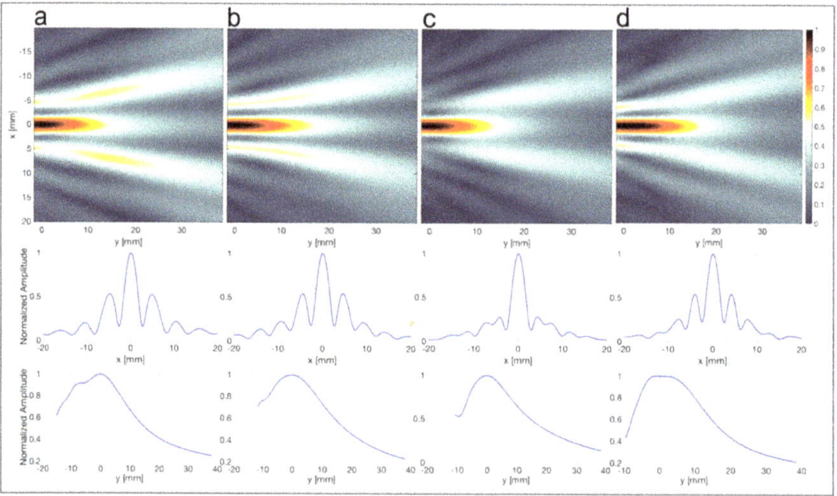

Figure 8. 445 kHz-Cigar-shaped acoustic focus for different settings of the stand-off. (**a**) δ = 0 mm; (**b**) δ = 20.5 mm; (**c**) δ = 34 mm; (**d**) δ = 45.25 mm. (Top) Focusing behavior of the PDMS-based 144° axicon lens with four different values of δ. (Middle) Normalized pressure amplitudes of the lateral beam profile. (Bottom) Normalized pressure amplitudes of the axial beam profile, 0 mm indicates the focus.

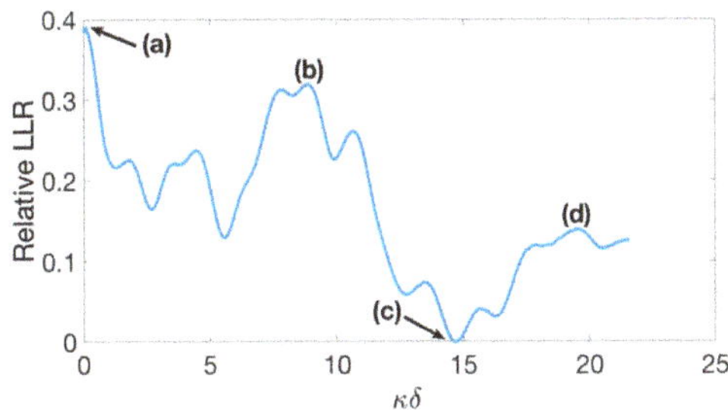

Figure 9. Stand-off, given in number of wavelengths, vs. relative loss of lateral resolution for 445 kHz PDMS-based 144° axicon lens. (**a**) $\delta = 0$ mm; (**b**) $\delta = 20.5$ mm; (**c**) $\delta = 34$ mm; and (**d**) $\delta = 45.25$ mm.

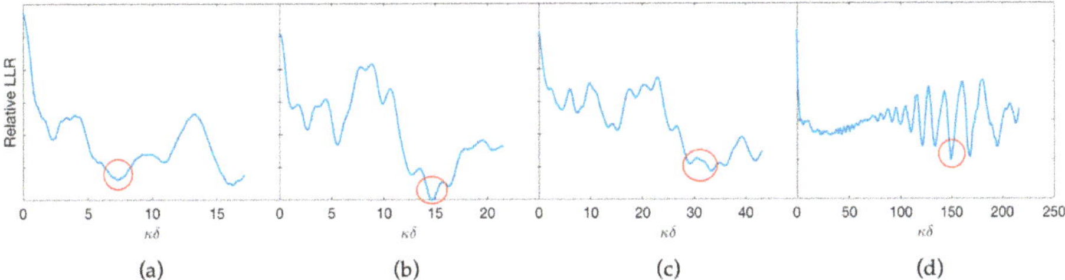

Figure 10. Stand-off given in number of wavelengths vs. relative loss of lateral resolution for different frequencies. The value of $\kappa\delta$ for the optimum stand-off (red circle) is directly proportional to the frequency. (**a**) f = 0.2225 MHz; (**b**) f = 0.445 MHz; (**c**) f = 0.890 MHz; (**d**) f = 4.45 MHz.

3.4. Acoustic-Field Experimental Scan

The hydrophone scans were performed both without and through the phantom. For the scan with the skull, the starting distance to the transducer was increased to 10 mm to avoid collision between the skull and hydrophone. Experimental beam patterns produced are shown in Figure 11. The lateral dimension of FUS beam cross-profiles measured at −6 dB drop of the pressure at the focus was 3.5 mm in the free space condition and 5 mm after transcranial transmission. We also characterized the acoustic field in the axial direction, perpendicular to the lens face and skull. The FUS pressure half width of the half maximum was 22 mm in the free space condition and 18 mm after transcranial transmission. Under these conditions, transmission of 445 kHz FUS-axicon lens through the skull led to an approximately 40% loss in lateral resolution of the acoustic beam, and on the other hand, an approximately 18% increase in the axial resolution. When FUS was transmitted through the skull, the acoustic pressure dropped by half. The insertion loss of our skull phantom was approximately −6 dB.

Intracranial focal characteristics obtained by simulation using the same configuration of Table 2 for different thicknesses of the skull are indicated in Table 3. There is a good coincidence between the simulation and the values obtained experimentally in water for the skull thickness of 5 mm, separated 2 mm from the lens, as shown in Table 4.

Table 3. Influence of the thickness skull in the focus of axicon lenses.

Thickness Skull (ts)	0.75 mm ts < λ/4	1.25 mm ts < λ/4	1.5 mm λ/4 < ts < λ/2	2.5 mm λ/4 < ts < λ/2	3 mm λ/2 < ts < λ	5 mm λ/2 < ts < λ	6 mm ts > λ
F (mm)	10.75	11	11	9.75	8.25	9.5	10.5
d_F (mm)	3.5	3.5	3.5	5	4	4	4
DOF (mm)	17	18	18	17.5	17	17.5	17
SLL (dB)	−8.2	−10.3	−11.5	−8.4	−8.7	−9.9	−9.5

Table 4. Comparison of simulated and scanned acoustic beam properties obtained using a configuration of 445 kHz transducer with a 144° Epoxy/PDMS lens, through the phantom of 5 mm thickness, separated 2 mm from the lens.

Beam Properties	Simulated	Scanned
d_F (mm)	4	5
DOF (mm)	17.5	18
SLL (dB)	−9.9	−14.4

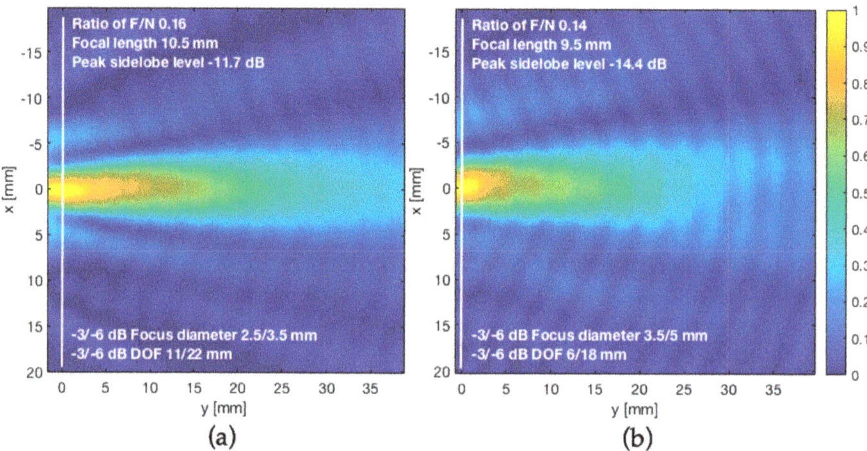

Figure 11. Ultrasound can be focused through human skull phantom. Experimental measurements of acoustic pressure field emitted from a 445 kHz axicon lens. (**a**) Free space without skull; (**b**) after transcranial transmission through skull bone phantom. White line indicates the focus.

4. Discussion

Although the analysis was carried out with a single element transducer with no aberration correction, and a specific skull geometry designed to approximate the varied shaped of the skull, it is expected that the relative influence of different medium properties and aspects of medium geometry will be maintained. Phantom geometry is the major material influence on the intracranial field and sound speed is shown to be the most influential acoustic property in focus pressure, position, and volume. From the experimental beam patterns shown in Figure 11, the unexpected focusing property of the skull in axial axis may be described as a nonlinear effect that causes the beam to rotate back toward the skull insertion point, creating a more compact pressure cigar-shaped acoustic field. This was also observed using segmented-sphere transducers [16,17]. Thus, the skull is not an obstacle for transcranial focusing of US and may exert an additional acoustic lensing effect to enhance spatial resolution under certain conditions. Out of focus, the sound pressure decreases with a very steep slope. From the comparison in Table 4, there is a good coincidence between the simulation and the values obtained experimentally in water for skull thickness of 5 mm, separated 2 mm from the lens. Regarding the influence of the thickness skull in the focus of axicon lenses, since on both sides of the skull there is a discontinuity in the acoustic impedance, with different velocities of sound propagation, and the average sound speed is the acoustic property that most influences the focal distance, we observe as expected, small variations in the position, diameter, and depth of focus, for different thicknesses of the cranial bone, resulting in the average deviations of the focus less than 1 mm.

On the other hand, transmitting FUS through human cranial bone caused an approximately 40% loss in lateral resolution of the acoustic beam, estimated by the intensity full width at half maximum. However, this loss of resolution is compensated by adequate stand-off of axicon lens. In addition, we find that the PDMS provides a smaller focal zone, which is desirable for neurostimulation [6]. All this allows a higher resolution, comparable to spherical transducers (the most commonly used for ultrasonic brain therapy), but with the advantage that the near field is eliminated, and the focus distance is shortened. For other applications, a wider focal zone and a line focus such as that obtained with glycerin or ethylene glycol may be desirable [7]. From Equation (5), the elimination of the near field, by means of an appropriate axicon lens, enables transducers featuring the same wavelength/diameter ratio to produce the same focal spot size. Thus, large diameter, low-frequency transducers may be used. This is useful for brain stimulation where low frequencies are required for penetration of the skull.

One problem of devices with axicon lenses are the relatively high sidelobes [18]. How much this will affect will depend on the proposed applications. With the calculation of the optimum value of δ, by Equation (6), a better lateral resolution is achieved. This relation between axicon lens stand-off and the value of F/N is applicable to high-resolution epoxy resin/PDMS lenses or other similar combination.

5. Conclusions

The numerical approach proposed in this paper provides a complete and effective way of designing axicon lenses for many high-resolution applications, such as mapping or detection. This is also suitable for focused ultrasound through human skull bone. In view of providing better lateral resolution with lower sidelobes, the use of design programs for this task is not as straightforward. The choice of a good starting point is an important factor for successful optimization. It is easier to obtain a starting design using relatively simple formulas and then use it in a lens design program for future analysis and optimization. We believe that this will be an effective way of designing axicon lenses, for example, to build focused windows to the brain for clinically viable transparent cranial implant for chronic ultrasonic therapy and stimulation of the brain.

Author Contributions: Conceptualization, F.A., S.E.L., and S.N.G.; data curation, F.A. and S.N.G.; formal analysis, F.A.; methodology, F.A. and S.N.G.; project administration, S.E.L.; software, F.A. and S.N.G.; supervision, S.E.L.; validation, F.A.; writing—original draft, F.A.; writing—review and editing, S.E.L. and S.N.G.

Funding: This research received no external funding.

Conflicts of Interest: The authors declare no conflict of interest.

Appendix A k-Wave Simulations

The simulations are based on the k-space pseudo-spectral method and implemented in MATLAB® with the open-source k-Wave toolbox. The functions based on the coupled first-order acoustic equations (named kspaceFirstOrder2D) are named with four input structures (kgrid, medium, source, and sensor). These structures define the properties of the computational grid, the distribution of medium properties, stress and velocity source terms, and the locations of the sensor points used to record the evolution of the wave field over time. The propagation of the wave field is computed step by step in a 2-D layered medium, with the pressure values at the sensor elements stored after each iteration. These values are returned when the time loop has completed. A list of the main simulation inputs is given in Table A1. One of the advantages of k-Wave is that the spatial gradients are calculated by using FFTs rather than using a finite-difference stencil. This means that for linear simulations, only two points per wavelength are required (Nyquist). For nonlinear simulations, the number of points per wavelength at the fundamental frequency will depend on the highest frequency of interest. The amplitudes of the harmonics should decay smoothly.

Table A1. Summary of the simulation inputs for relationships estimation of the axicon lens angle and the optimum stand-off.

Field	Value
Grid	
Number of grid points (N_X)	512
Size of the domain (x)	128×10^{-3} m
Grid point spacing ($d_X = x/N_X$)	2.5×10^{-4} m
Number of grid points for the sensor mask	512×512
Perfect Match Layer thickness	20 grid points
Transducer	
Sinusoidal transducer frequency (f)	0.515×10^6 Hz
Radius of the disc transducer (D/2)	28×10^{-3} m
Target medium (water)	
Sound speed (c)	1480 m/s
Density (ρ)	1000 kg/m^3
Stand-off and filling medium of the Axicon lens cavity (PDMS)	
Sound speed (c_1)	1030 m/s
Density (ρ_1)	1030 kg/m^3
Axicon lens cavity medium (epoxy resin)	
Sound speed (c_2)	2530 m/s
Density (ρ_2)	1170 kg/m^3
Inner sleeve	
Echo Reduction (ER)	-40 dB

Initially, the computational grid is defined using the function makeGrid. The time steps used in the simulation are defined by the object property kgrid.t_array. The time array was defined using the function makeTime, which calculate within the simulation functions using the time taken to travel across the longest grid diagonal at the slowest sound speed, and a Courant–Friedrichs–Lewy (CFL) number of 0.1, where CFL = $c_0 \Delta t / \Delta x$. The computational grid is defined by:

```
% size of the computational grid
Nx = 512; % number of grid points in the x (row) direction
x = 128e-3; % size of the domain in the x direction [m]
dx = x/Nx; % grid point spacing in the x direction [m]
% create the computational grid
kgrid = makeGrid(Nx, dx, Nx, dx);
% create the time array
[kgrid.t_array, dt] = makeTime(kgrid, medium.sound_speed);
```

After the computational grid, the properties of the propagation medium shown in Figure A1 are defined by the objects medium.sound_speed and medium.density. Also, the object medium.BonA represents the nonlinear properties of the medium. With medium.BonA = 0, k-Wave will include convective nonlinear effects in the model, but not include the nonlinear relationship between the acoustic pressure and acoustic density. If medium.BonA is undefined, k-Wave instead solves linearized equations.

The parameters medium.alpha_coeff (a_0) and medium.alpha_power (y) describe the power law acoustic attenuation in the medium, where the attenuation is of the form $a = a_0 \times f^\hat{} y$. The power law absorption and acoustic nonlinearity of water is specified by:

medium.alpha_power = 2; %[dB/(MHz^y cm)]
medium.alpha_coeff = 2.17e-3; %[dB/(MHz^y cm)]
medium.BonA = 4.96;

Next, a time varying pressure source is defined by assigning a binary source mask to source.p_mask (which defines the position of the source points) along with a time varying source input to source.p. Here, a single sinusoidal time series is used to drive the transducer element. The pressure source is defined by:

```
% define a transducer element
source.p_mask = makeLine(Nx, Ny, startpoint, pi/2, 112);
% define a time varying sinusoidal source
source_freq = 0.515e6; % [Hz]
source_mag = 1.0; % [Pa]
source.p = source_mag*sin(2*pi*source_freq*kgrid.t_array);
% filter the source to remove any high frequencies not supported %by the grid
source.p = filterTimeSeries(kgrid, medium, source.p);
```

Finally, to visualize the acoustic beam produced by the axicon lens, a sensor mask covering the entire computational domain was defined. Only the total beam pattern is required, thus at each time step, k-Wave only updates the maximum and r.m.s. values of the pressure at each sensor point by setting sensor.record to {'p_final', 'p_max', 'p_rms'}. The sensor mask is given here:

```
% create a sensor mask covering the entire computational domain
% using the opposing corners of a rectangle
sensor.mask = ones(Nx, Ny);
% set the record mode to capture the final wave-field and the
% statistics at each sensor point
sensor.record = {'p_final', 'p_max', 'p_rms'};
```

When the input structures have been defined, the simulation is started by passing them to kspaceFirstOrder2D.

The simulation is invoked by:

```
% create a display mask to display the transducer
display_mask = source.p_mask;
% assign the input options
input_args = {'DisplayMask', display_mask, 'PMLInside', false, ... 'DataCast', 'gpuArray-single', 'PlotPML', true, 'PlotLayout', true};
% run the simulation
sensor_data = kspaceFirstOrder2D(kgrid, medium, source, sensor, ... input_args{:});
```

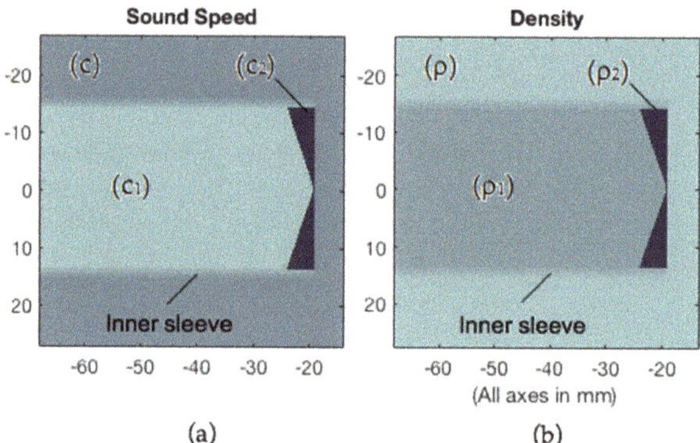

Figure A1. (a) Sound speed mask for the different mediums; (b) density mask for the different mediums.

The computed focal length (F) and optimum stand-off (δ) values for different lens angles (ϕ) of epoxy and PDMS axicon, with a transducer near-field distance in the water of 68.2 mm, are shown in Table A2.

Table A2. Values, obtained by running a 512 × 512 simulation, of focal length (F), ratio of F/N, optimum stand-off ($\delta_{Optimum}$), focus diameter (d_F), and maximum improvement of lateral resolution (MILR), for epoxy/PDMS-515 kHz Ø28mm axicon lens with different angles.

Angle ϕ	F (mm)	F/N (N = 68.2 mm)	$\delta_{Optimum}$ (mm)	$\kappa\delta_{Optimum}$ (λ = 2 mm)	d_F (mm @-6dB)	MILR (%)
115°	0.25	0.0037	29.0	14.50	1.50	65
120°	0.50	0.0073	32.0	16.00	2.00	58
125°	2.25	0.033	37.0	18.50	2.50	50
130°	3.75	0.055	38.5	19.25	2.50	51
135°	6.25	0.092	44.5	22.25	2.50	48
140°	9.00	0.13	45.0	22.50	3.00	54
145°	11.00	0.16	45.0	22.50	3.00	49
150°	14.50	0.21	45.5	22.75	3.50	51
155°	19.25	0.28	51.0	25.50	4.00	49
160°	22.75	0.33	51.0	25.50	4.50	47
165°	28.75	0.42	52.5	26.25	5.50	45
170°	33.00	0.48	53.0	26.50	6.00	44
175°	42.50	0.62	52.0	26.00	8.00	51

Appendix B Phantom Design

The Skull phantoms were created for the experimental validation using rapid prototyping techniques. The phantom was based on a 3D mesh of the parietal portion of the human skull. This was derived from a human body polygon dataset called "BodyParts3D" [19]. BodyParts3D is maintained by the Database Center for Life Science research located at the University of Tokyo. Polygon data are extracted from full-body MRI images. The MRI image set that BodyParts3D is based on is called "TARO". TARO is a 2 mm × 2 mm × 2 mm voxel dataset of the human male created by the National Institute of Information and Communications Technology [20]. BodyParts3D polygon data are distributed in the OBJ file format. The 3D mesh was segmented and smoothed with Meshmixer™.

The phantom was 3D printed in Clear Med610, with a one-layered homogeneous structure, using a Connex500 polyjet printer (Stratasys™). Although scattering due to the porous structure of the real skull could be expected to reduce focused transmission (on the other hand, it is possible that the random scattering reduces too existing destructive interference effects), the wavelength corresponding to ultrasound frequency used for 445 kHz is 3.37 mm in water, so that for this frequency, the dimensions of the skull inclusions are smaller than one-half the wavelength and, therefore, ultrasound will not be severely scattered by these inclusions.

Appendix C Material Characterization

Clear Med610 material was used to create the skull bone phantom described in Section 2 and Appendix B. The acoustic properties of Clear Med610 are equal to those of the VeroBlack that were reported in [15]; thus, these measurements were not repeated. However, independent measurements of the sound speed and attenuation of epoxy resin and PDMS were conducted and are shown in Table A3. To obtain reliable simulations, it is very important to accurately know the propagation velocity of ultrasound through the materials used for the lens. For materials characterization, both were poured into cylindrical molds and left to set at 25 °C for 48 h. The velocity of sound in epoxy resin and PDMS samples was determined by time of flight technique using an ultrasonic echoscope Digital-Echograph 1090 of Karl Deutsch. This measurement was performed in reflection mode with 2 MHz probe Karl Deutsch S6WB2.25. The sound speed in these materials has a weak dependence on frequency, less than 1% measured from 500 kHz to 2 MHz. The Attenuation of ultrasound was determined in reflection at the applied frequency of 2 MHz. The dimensions and weight of the test samples were measured by a micrometer and a digital scale, respectively.

Table A3. Acoustic properties of materials, distilled water, and tissue.

Material	Velocity (mm/µs@2MHz)	Density (g/cm³)	Impedance (MRayl)	Loss (dB/cm@2MHz)
Epoxy (25 °C)	2.53	1.17	2.96	6.8
PDMS (25 °C)	1.03	1.03	1.06	5.3
Water (20 °C)	1.48	1.00	1.48	0.08

Appendix D Guideline for Axicon Lens Design

This guide is based on design Equations (4)–(6) for Epoxy/PDMS materials combination.

(1) Calculate transducer near field, as:

$$N = \frac{D^2 f}{4c}$$

where D is transducer diameter, f is transducer frequency, and c is sound velocity of the material under inspection

(2) Select the desired lens focus F and check that:

$$0.1 \leq \frac{F}{N} < 0.4$$

Values of 0.1 are rarely used because it gives a very near focus. Values or 0.4 give a profile similar to the transducer but remove the N zone.

(3) Calculate the angle of the axicon lens, as:

$$\phi = \frac{9.9708 + \ln\left(\frac{F}{N}\right)}{5.52 \cdot 10^{-2}} \ [degrees].$$

(4) Calculate the optimum value of stand-off, as:

$$\delta_{Optimum} = \frac{10.921 + \ln\left(\frac{F}{N}\right)}{1.96 \cdot 10^2} \ [meters].$$

(5)   Focus diameter is:

$$d_F = \frac{D \cdot F}{2N}.$$

(6)   Depth of focus is:

$$DOF_F = 2F.$$

Figure A2 shows the relative LLR as a function of $\kappa\delta$ for value of F/N between 0.1 and 0.3.

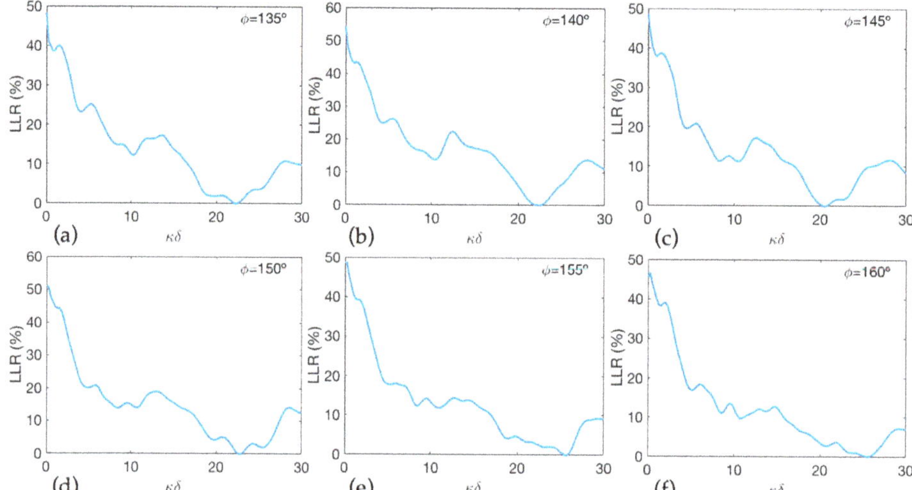

Figure A2. The relative percentage loss of lateral resolution (LLR) as a function of $\kappa\delta$. (**a**) F/N = 0.092; (**b**) F/N = 0.13; (**c**) F/N = 0.16; (**d**) F/N = 0.21; (**e**) F/N = 0.28; (**f**) F/N = 0.33.

References

1.   Mehić, E.; Xu, J.M.; Caler, C.J.; Coulson, N.K.; Moritz, C.T.; Mourad, P.D. Increased anatomical Specificity of neuromodulation via modulated focused ultrasound. *PLoS ONE* **2014**, *9*, e86939. [CrossRef] [PubMed]
2.   Robertson, J.L.; Cox, B.T.; Jaros, J.; Treeby, B.E. Accurate simulation of transcranial ultrasound propagation for ultrasonic neuromodulation and stimulation. *J. Acoust. Soc. Am.* **2017**, *141*, 1726–1738. [CrossRef] [PubMed]
3.   Yoo, S.S.; Bystritsky, A.; Lee, J.H.; Zhang, Y.; Fischer, K.; Min, B.K.; McDannold, N.J.; Pascual-Leone, A.; Jolesz, F.A. Focused ultrasound modulates region-specific brain activity. *Neuroimage* **2011**, *56*, 1267–1275. [CrossRef] [PubMed]
4.   Rezayat, E.; Toostani, G. A review on brain stimulation using low intensity focused ultrasound. *Basic Clin. Neurosci.* **2016**, *7*, 187–194. [PubMed]
5.   Fry, F.J.; Ades, H.W.; Fry, W.J. Production of reversible changes in the central nervous system by ultrasound. *Science* **1958**, *127*, 83–84. [CrossRef] [PubMed]
6.   Acquaticci, F.; Guarracino, J.F.; Gwirc, S.N.; Lew, S.E. A polydimethylsiloxane-based axicon lens for focused ultrasonic brain stimulation techniques. *Acoust. Sci. Technol.* **2019**, *40*, 116–126. [CrossRef]
7.   Katchadjian, P.; Desimone, C.; Garcia, A.D. Application of axicon lenses in ultrasonic techniques. *AIP Conf. Proc.* **2010**, *1211*, 1043–1050.
8.   Murphy, R.V. Focussed ultrasonic probes for contact inspection. *Mater. Eval.* **1980**, *38*, 53–58.
9.   Treeby, B.E.; Jaros, J.; Rendell, A.P.; Cox, B.T. Modeling nonlinear ultrasound propagation in heterogeneous media with power law absorption using a k-space pseudospectral method. *J. Acoust. Soc. Am.* **2012**, *131*, 4324–4336. [CrossRef] [PubMed]
10.   Treeby, B.E.; Cox, B.T. k-Wave: MATLAB toolbox for the simulation and reconstruction of photoacoustic wave fields. *J. Biomed. Opt.* **2010**, *15*, 021314. [CrossRef] [PubMed]

11. Wang, K.; Teoh, E.; Jaros, J.; Treeby, B.E. Modelling Nonlinear Ultrasound Propagation in Absorbing Media using the K-Wave Toolbox: Experimental Validation. In Proceedings of the 2012 IEEE International Ultrasonics Symposium, Dresden, Germany, 7–10 October 2012; pp. 523–526.

12. Liu, Q.H. Large-scale simulations of electromagnetic and acoustic measurements using the pseudospectral time-domain (PSTD) algorithm. *IEEE Trans. Geosci. Electron.* **1999**, *37*, 917–926.

13. Caputo, M.; Carcione, J.M.; Cavallini, F. Wave simulation in biologic media based on the Kelvin-voigt fractional-derivative stress-strain relation. *Ultrasound Med. Biol.* **2011**, *37*, 996–1004. [CrossRef] [PubMed]

14. Meza-Fajardo, K.C.; Papageorgiou, A.S. On the stability of a non-convolutional perfectly matched layer for isotropic elastic media. *Soil Dyn. Earthq. Eng.* **2010**, *30*, 68–81. [CrossRef]

15. Robertson, J.; Martin, E.; Cox, B.; Treeby, B.E. Sensitivity of simulated transcranial ultrasound fields to acoustic medium property maps. *Phys. Med. Biol.* **2017**, *62*, 2559–2580. [CrossRef] [PubMed]

16. Legon, W.; Sato, T.F.; Opitz, A.; Mueller, J.; Barbour, A.; Williams, A.; Tyler, W.J. Transcranial focused ultrasound modulates the activity of primary somatosensory cortex in humans. *Nat. Neurosci.* **2014**, *17*, 322–329. [CrossRef] [PubMed]

17. Lee, W.; Kim, H.; Jung, Y.; Song, I.U.; Chung, Y.A.; Yoo, S.S. Image-guided transcranial focused ultrasound stimulates human primary somatosensory cortex. *Sci. Rep.* **2015**, *5*, 8743. [CrossRef] [PubMed]

18. Burckhardt, C.B.; Hoffmann, H.; Grandchamp, P.-A. Ultrasound Axicon: A Device for Focussing Over Large Depth. *J. Opt. Soc. Am.* **1973**, *54*, 1628–1630. [CrossRef]

19. Mitsuhashi, N.; Fujieda, K.; Tamura, T.; Kawamoto, S.; Takagi, T.; Okubo, K. BodyParts3D: 3D structure database for anatomical concepts. *Nucleic Acids Res.* **2009**, *37*, 782–785. [CrossRef] [PubMed]

20. Nagaoka, T.; Watanabe, S.; Sakurai, K.; Kunieda, E.; Watanabe, S.; Taki, M.; Yamanaka, Y. Development of realistic high-resolution whole-body voxel models of Japanese adult males and females of average height and weight, and application of models to radio-frequency electromagnetic-field dosimetry. *Phys. Med. Biol.* **2004**, *49*, 1. [CrossRef] [PubMed]

Article

Effects of Loading and Boundary Conditions on the Performance of Ultrasound Compressional Viscoelastography: A Computational Simulation Study to Guide Experimental Design

Che-Yu Lin [1],* and Ke-Vin Chang [2,3],†

[1] Institute of Applied Mechanics, College of Engineering, National Taiwan University, No. 1, Sec. 4, Roosevelt Road, Taipei 10617, Taiwan
[2] Department of Physical Medicine and Rehabilitation, National Taiwan University Hospital and National Taiwan University College of Medicine, Taipei 100, Taiwan; kvchang011@gmail.com
[3] Department of Physical Medicine and Rehabilitation, National Taiwan University Hospital, Bei-Hu Branch, Taipei 10845, Taiwan
* Correspondence: cheyu@ntu.edu.tw
† Co-corresponding author.

Citation: Lin, C.-Y.; Chang, K.-V. Effects of Loading and Boundary Conditions on the Performance of Ultrasound Compressional Viscoelastography: A Computational Simulation Study to Guide Experimental Design. *Materials* **2021**, *14*, 2590. https://doi.org/10.3390/ma14102590

Academic Editor: Francesca Lionetto

Received: 8 March 2021
Accepted: 12 May 2021
Published: 16 May 2021

Publisher's Note: MDPI stays neutral with regard to jurisdictional claims in published maps and institutional affiliations.

Abstract: Most biomaterials and tissues are viscoelastic; thus, evaluating viscoelastic properties is important for numerous biomedical applications. Compressional viscoelastography is an ultrasound imaging technique used for measuring the viscoelastic properties of biomaterials and tissues. It analyzes the creep behavior of a material under an external mechanical compression. The aim of this study is to use finite element analysis to investigate how loading conditions (the distribution of the applied compressional pressure on the surface of the sample) and boundary conditions (the fixation method used to stabilize the sample) can affect the measurement accuracy of compressional viscoelastography. The results show that loading and boundary conditions in computational simulations of compressional viscoelastography can severely affect the measurement accuracy of the viscoelastic properties of materials. The measurement can only be accurate if the compressional pressure is exerted on the entire top surface of the sample, as well as if the bottom of the sample is fixed only along the vertical direction. These findings imply that, in an experimental validation study, the phantom design should take into account that the surface area of the pressure plate must be equal to or larger than that of the top surface of the sample, and the sample should be placed directly on the testing platform without any fixation (such as a sample container). The findings indicate that when applying compressional viscoelastography to real tissues in vivo, consideration should be given to the representative loading and boundary conditions. The findings of the present simulation study will provide a reference for experimental phantom designs regarding loading and boundary conditions, as well as guidance towards validating the experimental results of compressional viscoelastography.

Keywords: elastography; mechanical properties; viscoelastic properties; creep; stress relaxation

1. Introduction

Ultrasound elastography is an ultrasound-based imaging method for noninvasively measuring parameters related to the stiffness of materials [1–3]. This imaging technology was first described in the early 1990s [4] and subsequently developed into a real-time method for obtaining the map of parameters related to the stiffness of materials [1]. Ultrasound elastography is a clinical technique used to diagnose pathological conditions of various tissues [5], such as breast [6], liver [7], prostate [8], thyroid [9], tendons [10], muscles [11], and heel pads [12–14]. In addition, ultrasound elastography can potentially be used to evaluate the properties of a biomaterial for monitoring its development to ensure its quality [15–17]. It is based on the fact that pathological processes often cause changes in the stiffness of tissues [18,19], and the properties of a biomaterial are often related to its stiffness [15,20]. However, most tissues and biomaterials are viscoelastic [15]. This means that they have both viscous (fluid) and elastic (solid) properties [21]. Changes in the status of tissues and biomaterials often lead to alterations in both the fluid and solid properties. Therefore, it is important to characterize the viscoelastic properties if we intend to completely evaluate pathological tissues or the condition of an engineered biomaterial. In many circumstances, measuring the stiffness alone may not be sufficient to completely evaluate the status of tissues and biomaterials. Several studies have shown that viscosity is a better discriminator than stiffness to differentiate between malignant and benign tumors [22,23]. One study also suggested that considering viscosity can provide additional important information, rather than just considering stiffness alone [24].

In a laboratory setting, the viscoelastic properties of materials are often evaluated by mechanical material testing systems. Several ultrasound techniques have been designated by different research groups to evaluate the viscoelastic properties of materials based on analyzing the creep behavior (increasing strain over time, Figure 1), generally called viscoelastic creep imaging [25]. In viscoelastic creep imaging, the creep behavior of an element inside the material can be acquired by applying a constant stress to the material. The constant stress can be induced by ultrasound acoustic radiation force [16,17,26–32], or by external mechanical compression on the top surface of the material [33,34]. If a viscoelastic mathematical model is used to curve-fit the creep behavior, the viscoelastic properties of materials can be quantitatively evaluated. In the design and experimental setup of viscoelastic creep imaging, several factors should be carefully considered to achieve the best measurement accuracy and optimal performance of the system.

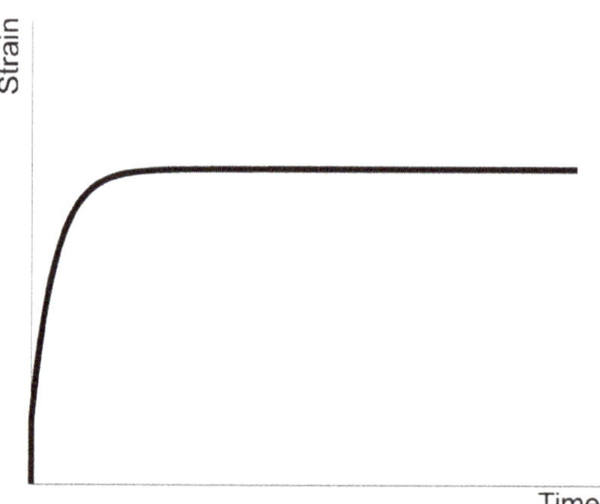

Figure 1. Illustration of creep behavior (increasing strain over time).

The present study will investigate the measurement accuracy of compressional viscoelastography (a type of viscoelastic creep imaging using external mechanical compression as the source of excitation) through computational simulations of imaging. The aim of the present study is to use finite element analysis to investigate the performance of compressional viscoelastography to measure the viscoelastic properties of homogeneous viscoelastic materials, and to investigate how loading conditions (the distribution of the applied compressional pressure on the surface of the sample) and boundary conditions (the fixation method used to stabilize the sample) can affect the measurement accuracy. The results of the present simulation study provide a reference for experimental phantom designs regarding loading and boundary conditions, as well as guidance towards validating the experimental results of compressional viscoelastography.

2. Materials and Methods

2.1. Principle of Compressional Viscoelastography

The design of the compressional viscoelastography system is shown in Figure 2 (refer to references [33,34] for a more detailed introduction of the design of this system). The system uses a circular pressure plate to apply a uniform compressional pressure on the surface of the material. The pressure and back plates are connected to each other and attached to the ultrasound transducer. The magnitude of the compressional pressure can be measured by load cells attached between the two plates. The system has a feedback control unit that allows for the simultaneous monitoring and control of the compressional pressure. The movement of the transducer is controlled by a stepper motor. The strain of an element within the material is measured using ultrasound imaging.

In the measurement, the transducer is moved downward to compress the material at a constant loading rate until the preset pressure level is reached. The loading rate must be high to mimic a step load. Once the preset pressure level is reached, the compressional pressure is maintained as constant through a feedback control system. During the period when the compressional pressure is maintained as constant, the stress of an element within the material is constant as well. Therefore, each element within the material exhibits creep behavior. The measurement processes are illustrated in Figure 3.

In the present study, the Maxwell representation of the standard linear solid model (hereafter referred to as the standard linear solid model for brevity), as shown in Figure 4, is used to describe the viscoelastic behaviors of materials. If a step stress excitation is used to excite the material, the creep behavior (increasing strain over time) described by the standard linear solid model is [35]:

$$\epsilon(t) = \frac{\sigma_0}{E_1}\left(1 - \frac{E_2}{E_1 + E_2} \cdot e^{-\frac{1}{\tau_C}t}\right) \tag{1}$$

where $\epsilon(t)$ is the strain over time of an element within the material; $\tau_C = \eta(E_1 + E_2)/E_1 E_2$ is the creep time constant (also called the retardation time constant); E_1, E_2, and η are parameters relevant to viscoelastic properties; σ_0 is the stress value of an element within the material following the step stress excitation at $t = 0$ (i.e., the beginning of creep); and σ_0 is the constant stress value during creep. In the present study, σ_0 in Equation (1) is assumed to be the magnitude of uniform compressional pressure applied on the surface of the material, although this assumption may not be perfectly accurate for each element within the material since the appropriateness of this assumption depends on the loading and boundary conditions.

If Equation (1) is used to curve-fit the creep curve of each element within the material, the E_1, E_2, and τ_C of each element can be quantitatively evaluated. Consequently, the spatial distribution of the viscoelastic properties of the material can be obtained, as described in the following section.

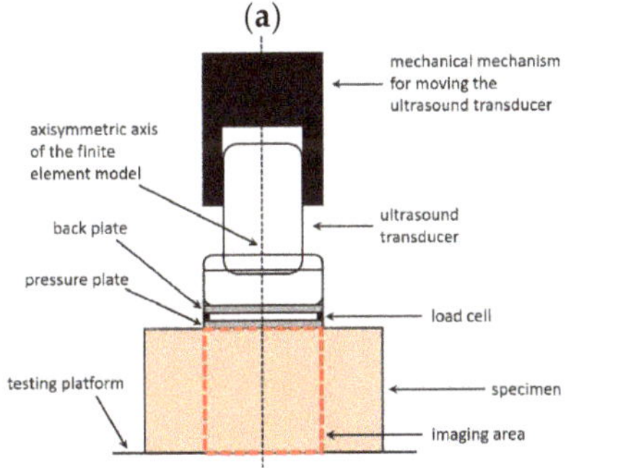

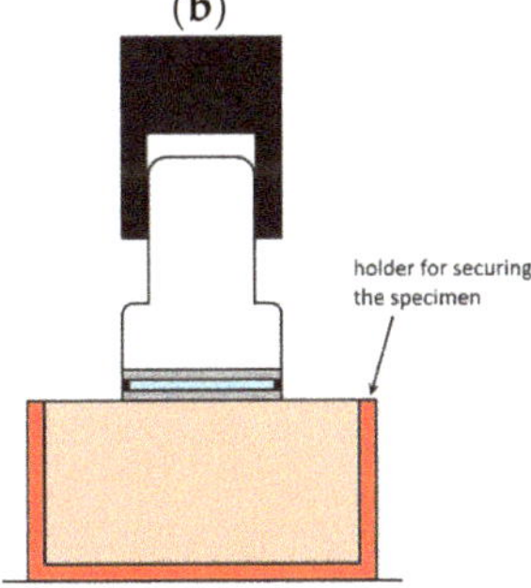

Figure 2. *Cont.*

Figure 2. The design of a compressional viscoelastography system with different experimental settings. (**a**) The surface area of the pressure plate is half of that of the top surface of the sample, and the sample is placed directly on the testing platform without any fixation. This setup is associated with the condition in Figure 5d. (**b**) The surface area of the pressure plate is half of that of the top surface of the sample, and the sample is secured by a sample container. This setup is associated with the condition in Figure 5f. (**c**) The surface area of the pressure plate is equal to that of the top surface of the sample, and the sample is placed directly on the testing platform without any fixation. This setup is associated with the condition in Figure 5a. (**d**) The surface area of the pressure plate is equal to that of the top surface of the sample, and the sample is secured by a sample container. This setup is associated with the condition in Figure 5c.

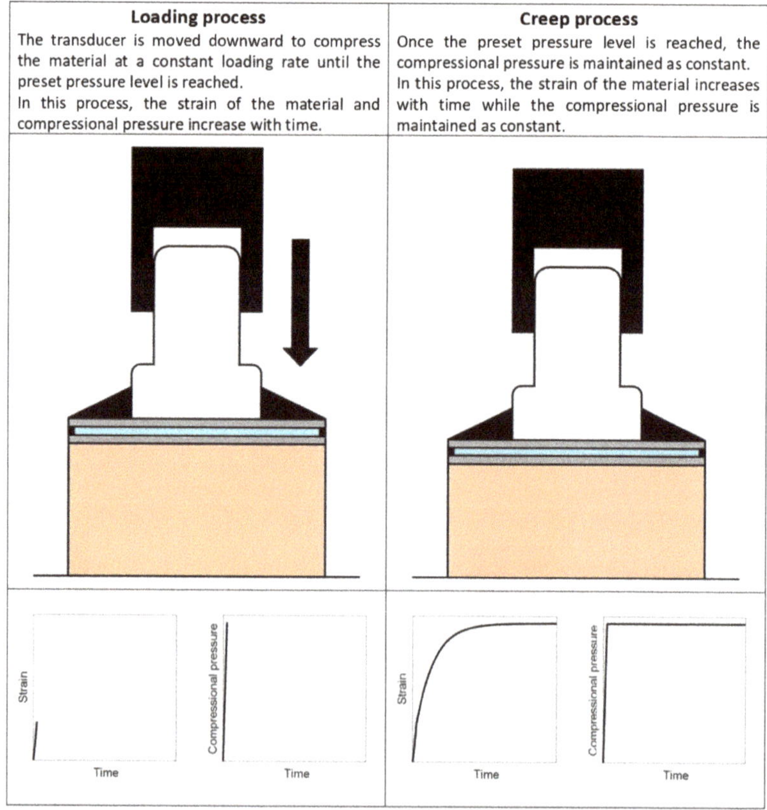

Figure 3. Illustration of the measurement processes of compressional viscoelastography.

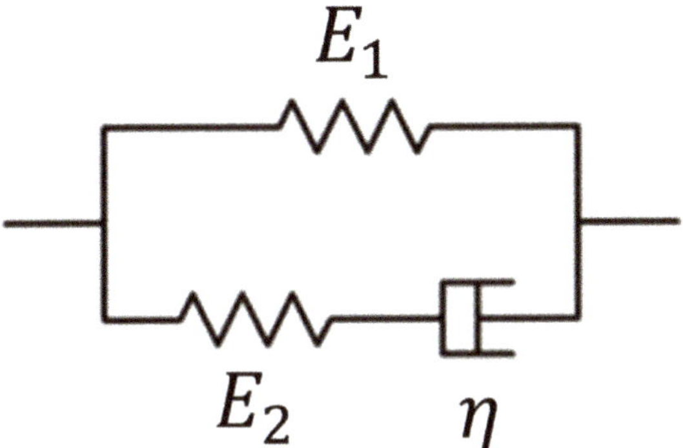

Figure 4. Illustration of the Maxwell representation of the standard linear solid model. E_1, E_2, and η are three viscoelastic properties.

2.2. Computational Simulations and Data Analysis

Finite element analysis using ABAQUS 2019 (Dassault Systems, Simulia Corporation, Johnson, RI, USA) is used to investigate the measurement accuracy of compressional viscoelastography when measuring the viscoelastic properties of the material under different loading and boundary conditions.

Six simulation tests were run in this study. The specific settings regarding the loading and boundary conditions in each simulation test are described below and illustrated in Figure 5:

(1)   Simulation test 1 (Figure 5a): uniform compressional pressure is exerted on the entire top surface. The bottom is fixed along the vertical direction while the side is not fixed. This condition is associated with the experimental setup in Figure 2c. Simulation test 1 is also used to investigate the validity of compressional viscoelastography. See Appendix A for the result of the validity test.

(2)   Simulation test 2 (Figure 5b): uniform compressional pressure is exerted on the entire top surface. The bottom is fixed along all directions while the side is not fixed.

(3)   Simulation test 3 (Figure 5c): uniform compressional pressure is exerted on the entire top surface. The bottom is fixed along the vertical direction while the side is fixed along the horizontal direction. This condition is associated with the experimental setup in Figure 2d.

(4)   Simulation test 4 (Figure 5d): uniform compressional pressure is exerted on half of the top surface. The bottom is fixed along the vertical direction while the side is not fixed. This condition is associated with the experimental setup in Figure 2a.

(5)   Simulation test 5 (Figure 5e): uniform compressional pressure is exerted on half of the top surface. The bottom is fixed along all directions while the side is not fixed.

(6)   Simulation test 6 (Figure 5f): uniform compressional pressure is exerted on half of the top surface. The bottom is fixed along the vertical direction while the side is fixed along the horizontal direction. This condition is associated with the experimental setup in Figure 2b.

The six simulation tests are different in their loading and boundary conditions as described above, but they have some common settings. In each simulation test, an axisymmetric finite element model is used, and the radius and thickness of the axisymmetric model are 50 mm and 50 mm, respectively. Square finite elements (0.5 mm × 0.5 mm) are used to mesh the model. The radius and thickness of the imaging region (the region of the model where data are taken to produce the image, which is chosen as the region of the model just below the transducer, with a length of 50 mm) are 25 mm and 50 mm, respectively, as shown in Figure 2a.

The model is made of linear viscoelastic material. The material is also assumed to be incompressible, isotropic, and homogeneous. The mechanical properties of the material are defined by four properties, including the modulus of elasticity (E), Poisson's ratio (set as a constant of 0.495, the maximum Poisson's ratio that can be set in ABAQUS), and two parameters in the one-branch dimensionless relaxation modulus:

$$g_R(t) = 1 - g\left(1 - e^{\frac{-t}{\tau_R}}\right) \tag{2}$$

where g is a material constant ($0 < g < 1$), and τ_R is the relaxation time constant.

The loading rate for applying the compressional pressure is designed to be high (the time duration from zero to maximum pressure is $1/6$ s) to mimic a step load. Once the maximum pressure (set as a constant of 1000 Pa in each simulation test) is reached, the maximum pressure is then maintained as a constant for a period of time. During this period, each element in the model exhibits creep behavior; that is, the strain response of each element increases with time until the steady state is reached. The creep curve of each element is sent to MATLAB (R2019a, Mathworks, Natick, MA, USA) for analysis. The E_1, E_2, and τ_C of each element can be obtained by using Equation (1) to curve-fit the creep curve of each element. η can then be obtained by using the definition of the creep time constant $\tau_C = \eta(E_1 + E_2)/E_1 E_2$. It has been reported that E is equal to E_1, while g is equal to $E_2/(E_1 + E_2)$, and τ_R is equal to η/E_2 [35]. Consequently, the three mechanical properties of each element (E, g, and τ_R) set in ABAQUS can be obtained.

Once the value of a mechanical property (E, g, or τ_R) of each element is assigned with a specific color, the 2D spatial distribution map of that mechanical property of the axisymmetric image region can be constructed (since the image region is quadrilateral and the elements within are square). However, this map of the axisymmetric image region is just half of the map that should be obtained in reality (since the model is axisymmetric). Therefore, this map of the axisymmetric image region (called the original map here) is reflected, and then combined with the original map to produce a full map.

The corresponding error map is also constructed for each mechanical property. The error value at each element in the error map is calculated by:

$$\text{error} = \frac{|\text{simulation value} - \text{theoretical value}|}{\text{theoretical value}} \tag{3}$$

where the simulation value is the value of the mechanical property of an element obtained from the simulation, and the theoretical value is the value of the mechanical property set in ABAQUS. If the error is less than 10% [36], the measurement is considered to be accurate.

In each simulation test, the mechanical properties are fixed and set as $E = 10$ kPa, $\tau_R = 5$ s, and $g = 0.8$ (because, in our preliminary study, it was found that the pattern of the map of a mechanical property is similar regardless of the mechanical properties; in other words, the pattern of the map of a mechanical property is sensitive to the loading and boundary conditions but insensitive to the mechanical properties). In the 2D spatial distribution map of each mechanical property in each simulation test, the percentage of the region in the map consisting of elements having the simulation value within $\pm 10\%$ [36] of the theoretical value set in ABAQUS is calculated. The larger this percentage, the more accurate the measurement, because the wider the region in the 2D spatial distribution map, the more accurate the value of the mechanical property.

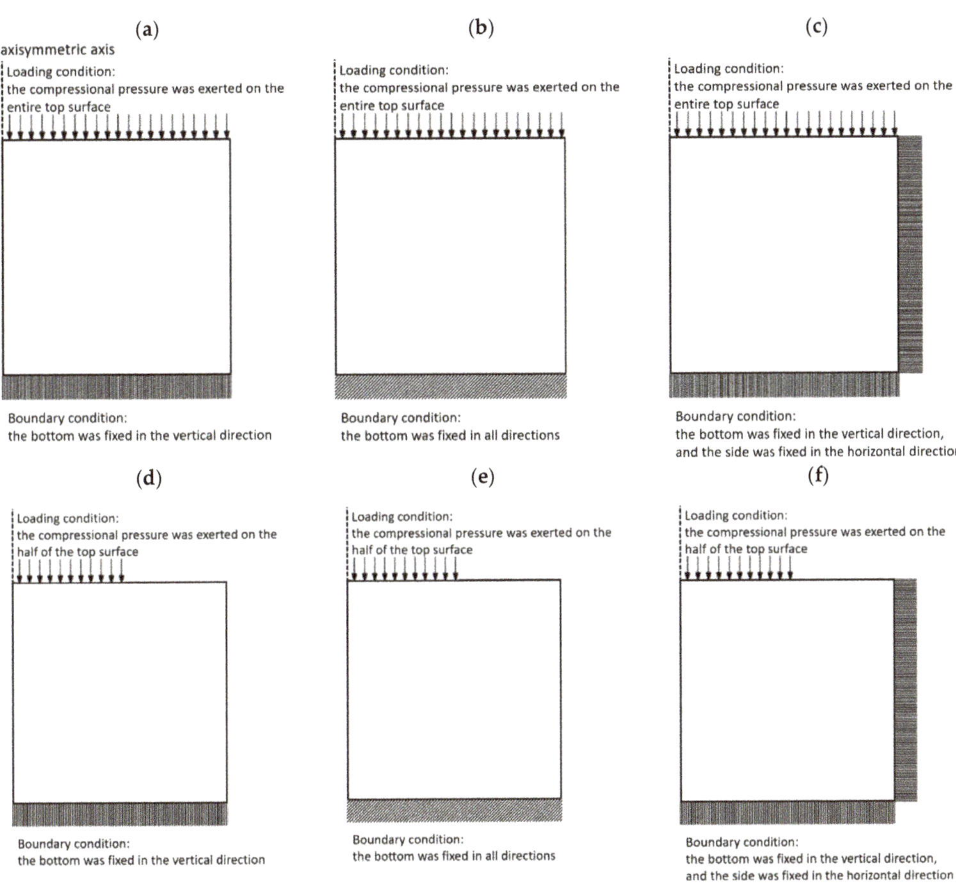

Figure 5. Illustration of the loading and boundary conditions in each simulation test. (**a**) The uniform compressional pressure is exerted on the entire top surface. The bottom is fixed along the vertical direction while the side is not fixed. (**b**) The uniform compressional pressure is exerted on the entire top surface. The bottom is fixed along all directions while the side is not fixed. (**c**) The uniform compressional pressure is exerted on the entire top surface. The bottom is fixed along the vertical direction while the side is fixed along the horizontal direction. (**d**) The uniform compressional pressure is exerted on half of the top surface. The bottom is fixed along the vertical direction while the side is not fixed. (**e**) The uniform compressional pressure is exerted on half of the top surface. The bottom is fixed along all directions while the side is not fixed. (**f**) The uniform compressional pressure is exerted on half of the top surface. The bottom is fixed along the vertical direction while the side is fixed along the horizontal direction.

3. Results

Figure 6 shows the 2D spatial distribution map and corresponding error map for each mechanical property in each simulation test. Table 1 shows the percentage of the region in the 2D spatial distribution map consisting of elements with a simulation value within ±10% of the theoretical value for each simulation test.

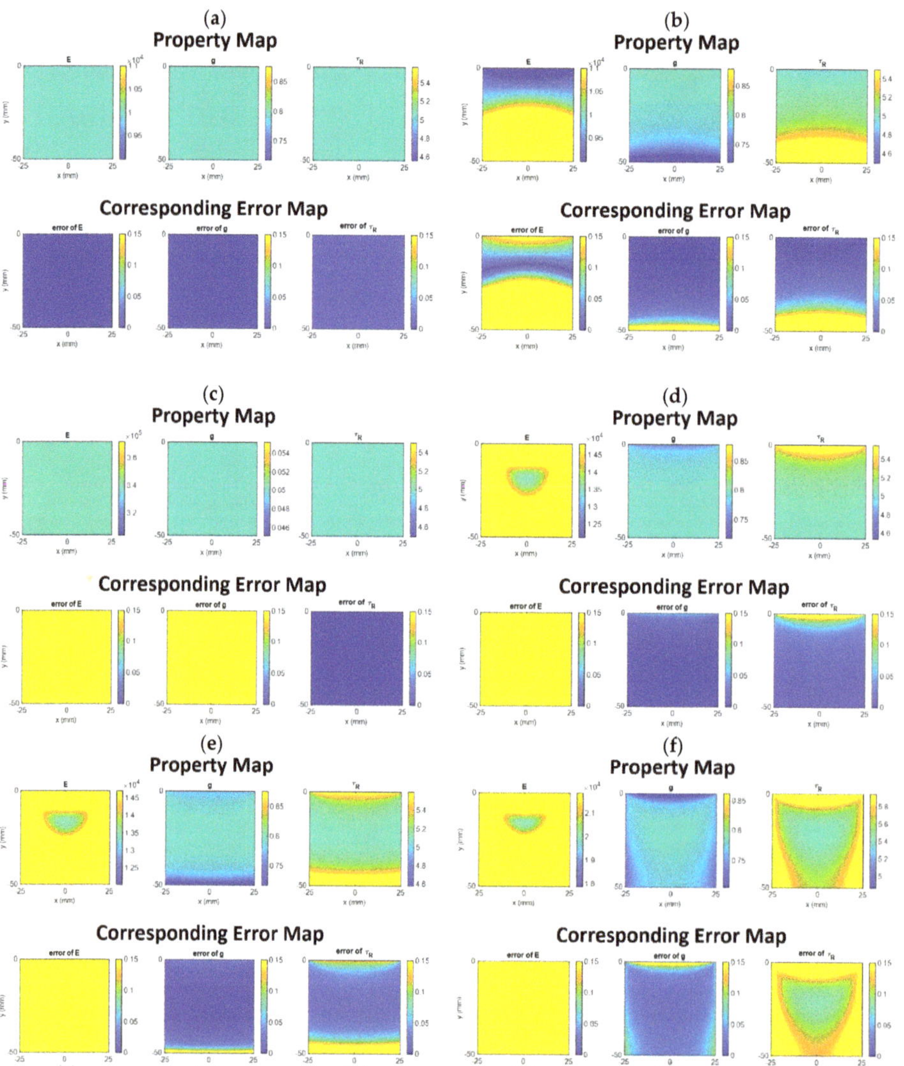

Figure 6. The 2D spatial distribution and corresponding error maps. In the error map, the yellow color means the error is larger than 10%. (**a**) Simulation test 1. (**b**) Simulation test 2. (**c**) Simulation test 3. (**d**) Simulation test 4. (**e**) Simulation test 5. (**f**) Simulation test 6.

Table 1. The percentage of the region in the map consisting of elements with the simulation value within ±10% of the theoretical value set in ABAQUS. The larger this percentage, the more accurate the measurement.

Simulation Test Number	Percentage of the Region in the Map Having Accurate Measurement (%)		
	E	τ_R	g
1	100	100	100
2	37.28	77.50	92.42
3	0	100	0
4	0	93.84	100
5	0	83.16	95.94
6	0	23.24	96.18

In the first simulation test (when the uniform compressional pressure is exerted on the entire top surface, and the bottom is fixed along the vertical direction while the side is not fixed), the 2D distribution map for each mechanical property is perfectly homogeneous, as is the corresponding error map. The error value at each element in the error map is nearly zero. This means that all three mechanical properties can be accurately measured in this case.

In the second simulation test (when the uniform compressional pressure is exerted on the entire top surface, and the bottom is fixed along all directions while the side is not fixed), g can be accurately measured in most of the region (92.42%), except for a small region close to the bottom. τ_R can be accurately measured in 77.5% of the entire region, except for a region extending from the bottom to near the depth of 40 mm. For E, it can only be accurately measured in 37.28% of the entire region, between the depth of 5 to 25 mm approximately, and there is a significant region where each element in this region has an error larger than 10%.

In the third simulation test (when the uniform compressional pressure is exerted on the entire top surface, and the bottom is fixed along the vertical direction while the side is fixed along the horizontal direction), the 2D distribution map for each mechanical property is perfectly homogeneous. However, only τ_R can be accurately measured, while E and g cannot be since each element in the entire region has an error larger than 10%.

In the fourth and fifth simulation tests (when the uniform compressional pressure is exerted on the half of the top surface, and the bottom is fixed along the vertical direction), g and τ_R can be accurately measured, except for a small region close to the top and/or bottom. E cannot be accurately measured since each element in the entire region has an error larger than 10%.

In the sixth simulation test (when the uniform compressional pressure is exerted on the half of the top surface, and the bottom is fixed along the vertical direction while the side is fixed along the horizontal direction), g can be accurately measured in most of the region (96.18%). For τ_R, it can only be accurately measured in 23.24% of the region. E cannot be accurately measured since each element in the entire region has an error larger than 10%.

4. Discussion

The findings of the present finite element analysis study provide a reference for experimental phantom designs regarding loading and boundary conditions, as well as guidance towards validating the experimental results of a compressional viscoelastography system, such as the one in the literature [33,34]. The most important finding is that the accuracy of compressional viscoelastography for measuring the viscoelastic properties of materials is only excellent if the compressional pressure is exerted on the entire top surface of the sample, as well as if the bottom of the sample is fixed just along the vertical direction. These findings imply that in an experimental validation study, the phantom design should take into account that the surface area of the pressure plate must be equal to or larger than that of the top surface of the sample, and the sample should be placed directly on the testing platform without any fixation (such as a sample container). However, in an experimental design, the sample may slip horizontally under loading if there is no fixation to stabilize the sample. Fortunately, the slip can be prevented by using a pressure plate with a surface area significantly larger than that of the top surface of the sample.

In the present study, it was found that the loading and boundary conditions in computational simulations of compressional viscoelastography severely affect the accuracy of measuring the modulus of elasticity, E. If the compressional pressure is exerted on half of the top surface of the sample and the boundary condition is more complex than the abovementioned optimal condition (the bottom of the sample is fixed just along the vertical direction), E is difficult measure accurately. This implies that if the area of the pressure plate is smaller than that of the top surface of the sample, and if the sample is secured by any external fixation method (such as the sample container shown in Figure 2b), E may not be accurately measured. In these kinds of conditions, there will be horizontal strains occurring within the sample. The theory used in the present study for analyzing the creep curve, Equation (1), is a one-dimensional model that only considers the strain component along the vertical direction. Therefore, if there are horizontal strains within the sample, Equation (1) may not yield precise measurements; the larger the horizontal strains, the greater the errors. This may be the reason why an accurate measurement for E cannot be achieved in the conditions where horizontal strains can develop within the sample. In addition, the distribution of horizontal strains affects the homogeneity of the 2D spatial distribution map of E. Figure 7 shows the distribution of horizontal strains in each simulation test. It can be observed that every test has horizontal strains. If the distribution of horizontal strains is homogeneous, such as that in the first and third tests, the 2D spatial distribution map of E is homogeneous. On the other hand, if the distribution of horizontal strains is nonhomogeneous, the 2D spatial distribution map of E is nonhomogeneous. Therefore, for an accurate measurement of E in more complex loading and boundary conditions, strain components along both horizontal and vertical directions must be considered simultaneously, and a multi-dimensional model must be used for data analysis. The measurements of τ_R and g are also significantly affected by loading and boundary conditions. It is interesting to note that when the compressional pressure is exerted on half of the top surface of the model, the measurement of g is less affected by the boundary condition. This means that g can always be accurately measured no matter what the boundary condition is. On the other hand, the more complex the boundary condition, the more difficult the measurement of τ_R.

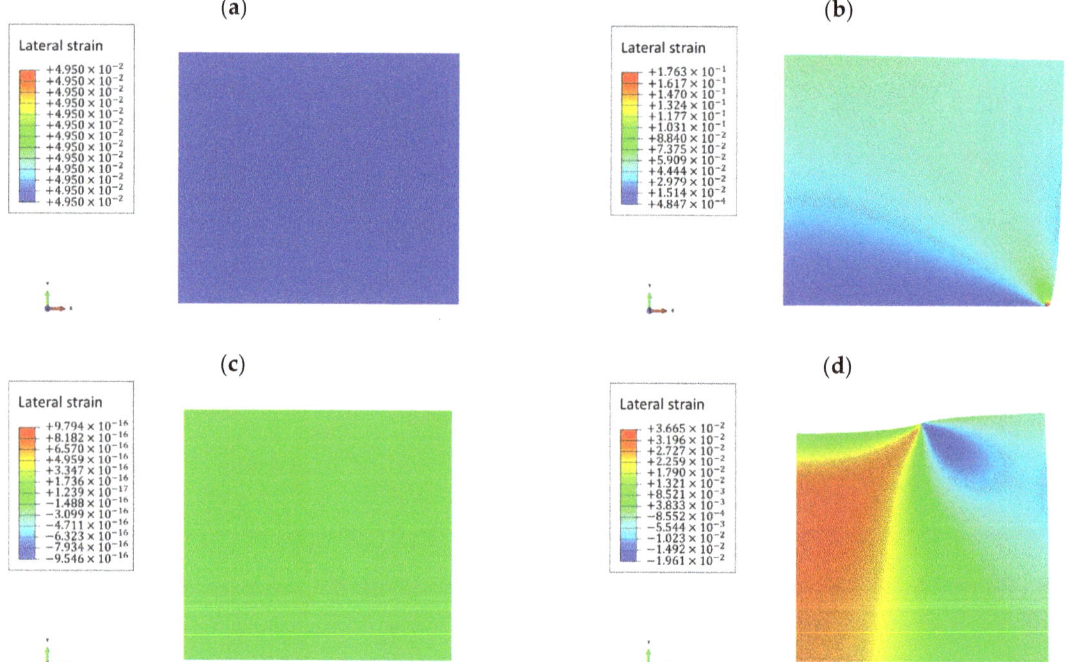

Figure 7. *Cont.*

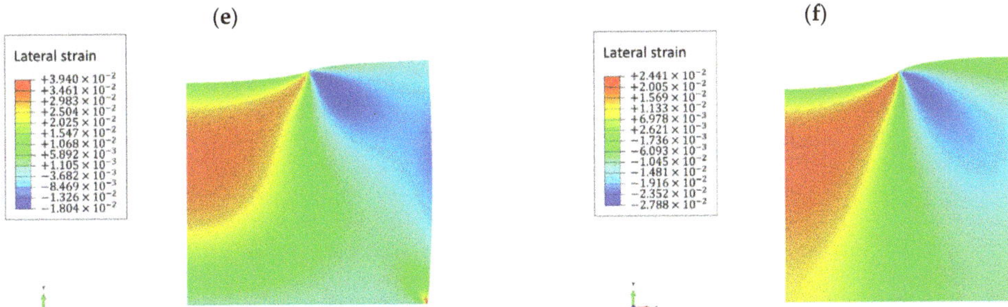

Figure 7. The distribution of horizontal strains in each simulation test. (**a**) Simulation test 1. (**b**) Simulation test 2. (**c**) Simulation test 3. (**d**) Simulation test 4. (**e**) Simulation test 5. (**f**) Simulation test 6.

The initial intention for developing compressional viscoelastography was to measure the viscoelastic properties of breast tissues in vivo for the diagnosis of breast tumors [33,34]. Unfortunately, in the present study, it has been found that the loading and boundary conditions in computational simulations of compressional viscoelastography severely affect the measurement accuracy. The measurement can only be accurate if the compressional pressure is exerted on the entire top surface of the sample, as well as if the bottom of the sample is fixed just along the vertical direction. However, the loading and boundary conditions could be much more complex than these optimal conditions when measuring real tissues in vivo. Therefore, the findings indicate that when applying compressional viscoelastography to real tissues in vivo, consideration should be given to the representative loading and boundary conditions. Further studies are needed to investigate the measurement accuracy of compressional viscoelastography on both biomaterials and real tissues in vivo.

Finite element analysis has been applied to investigate the performance or the effects of system parameters of ultrasound elastography [37–40]. There are also some valuable studies using finite element analysis to explore magnetic resonance elastography [41–44]. However, to the best of our knowledge, there are only two studies that have used finite element analysis to investigate the performance of ultrasound compressional viscoelastography on the measurement of the viscoelastic properties of materials [34,45]. It is the authors' belief that more studies are needed to investigate the performance of ultrasound compressional viscoelastography to justify its usefulness in clinical or biomedical applications.

Since the present study is based on finite element analysis, it is very important to further discuss and explain the simulation settings to ensure that the simulation results are understood and applied properly. First, the finite element model used in this study was a simple axisymmetric model. The findings could be successfully applied on biomaterials since they are often designed as an axisymmetric cylinder. In the future, a more complex shape for the model should be considered so that the findings can be more accurately applied to samples with different geometries. Second, the magnitude of applied force could be different across various simulation tests since the magnitude of compressional pressure was the same, but the area over which it was applied might vary across test. The reason for applying a constant pressure (but not a constant force) on each test is that the pressure can be regarded as the force normalized by the area over which it is applied. Therefore, by applying a constant pressure on each test, similar orders of magnitude of stress and strain can be induced within the model. Third, the volume of the model relative to the compression area is an important parameter, but this parameter was outside the scope of this study. During a compression test by compressional viscoelastography, if the aspect ratio (diameter/thickness) of the model is much smaller than one, buckling could occur. Therefore, in order to prevent buckling, the aspect ratio of the model must be larger than one. This is why the axisymmetric model was designed to have a constant volume with a radius of 50 mm and a thickness of 50 mm (the associated aspect ratio is two).

The present study has some further limitations: (1) not all variables relevant to compressional viscoelastography were investigated. In addition, no noise or imaging uncertainties were considered in the finite element analysis in this study; in reality, the signal of noise and vibration may affect the accuracy of the system. Therefore, the findings of this study only can provide suggestions and references for experimental phantom designs regarding loading and boundary conditions, but cannot provide global guidance for every technical aspect relevant to compressional viscoelastography. (2) Only two types of loading conditions were investigated. It will be valuable to conduct a more detailed parametric analysis to quantitatively investigate the relationship between the loading condition and the area of the image region with accurate measurements. (3) Only homogeneous materials were investigated, and an inclusion phantom was not considered. (4) There was no experiment to validate the finite element analysis results of this study. Both simulations and experiments are important and have their own merits and limitations. Simulation offers the possibility and convenience to exactly control a condition (for example, to set the viscoelastic properties of materials as specific values) for the analysis of a great variety of conditions, and can show a relative and general trend to provide guidance for experimental design. Experiments provide real data showing what happens in reality. In the future, it is important to investigate if the same results can be observed in real experiments.

5. Conclusions

In conclusion, the findings of the present simulation study will provide a reference for experimental phantom designs regarding loading and boundary conditions, as well as guidance towards validating the experimental results of compressional viscoelastography (viscoelastic creep imaging using external compression as the source of excitation). The results show that loading and boundary conditions in computational simulations of compressional viscoelastography can severely affect the measurement accuracy of the viscoelastic properties of materials. The measurement can only be accurate if the compressional pressure is exerted on the entire top surface of the sample, and if the bottom of the sample is fixed just along the vertical direction. These findings imply that, in an experimental validation study, the phantom design should take into account that the surface area of the pressure plate must be equal to or larger than that of the top surface of the sample, and the sample should be placed directly on the testing platform without any fixation (such as a sample container). These findings indicate that when applying compressional viscoelastography to real tissues in vivo, consideration should be given to the representative loading and boundary conditions.

Author Contributions: C.-Y.L. is responsible for the conceptualization, data curation, formal analysis, funding acquisition, investigation, methodology, software, visualization, and writing. K.-V.C. is responsible for the conceptualization, visualization, and writing. All authors have read and agreed to the published version of the manuscript.

Funding: The authors acknowledge the financial support provided by the Ministry of Science and Technology of Taiwan (grant number: MOST 108-2218-E-002-046-MY3).

Institutional Review Board Statement: Not applicable.

Informed Consent Statement: Not applicable.

Data Availability Statement: The datasets relevant to this study are available on request to the corresponding author.

Conflicts of Interest: The authors declare no conflict of interest.

Appendix A

Simulation test 1 was also used to investigate the validity of compressional viscoelastography. In this investigation, we studied three groups of materials with different mechanical properties (Table A1). In group 1, E = 10, 20, and 30 kPa, τ_R = 5 s, and g = 0.8.

In group 2, $\tau_R = 1$, 5, and 10 s, $E = 10$ kPa, and $g = 0.8$. In group 3, $g = 0.4$, 0.6, and 0.8, $E = 10$ kPa, and $\tau_R = 5$ s. For the map of each mechanical property, the simulation value (the median of the values of all elements) is compared to the theoretical value (the value of that mechanical property set in ABAQUS) using the following equation:

$$\text{error} = \frac{|\text{simulation value} - \text{theoretical value}|}{\text{theoretical value}}$$

Table A1 shows the results. It can be observed that the theoretical and simulation values are very close for each case. Figure 6a shows the 2D distribution map of each mechanical property and the corresponding error map for simulation test 1. It can be observed that the map for each mechanical property is perfectly homogeneous, as is the corresponding error map. The error for each element in the error map is nearly zero. The results show that the measurement is accurate, justifying the validity of compressional viscoelastography.

Table A1. The results of investigating the validity of compressional viscoelastography. It can be observed that the theoretical and simulation values are very close for each case.

Material Number	Theoretical Values			Simulation Values			Error (%)		
	E	τ_R	g	E	τ_R	g	E	τ_R	g
				Group 1					
1	10,000	5	0.8	10,001	5.067	0.797	0.01	1.35	0.34
2	20,000	5	0.8	20,003	5.068	0.797	0.01	1.37	0.34
3	30,000	5	0.8	30,006	5.069	0.797	0.02	1.38	0.34
				Group 2					
1	10,000	1	0.8	10,001	1.014	0.797	0.01	1.35	0.34
2	10,000	5	0.8	10,001	5.067	0.797	0.01	1.35	0.34
3	10,000	10	0.8	10,001	10.135	0.797	0.01	1.34	0.34
				Group 3					
1	10,000	5	0.4	10,000	5.011	0.399	0.00	0.23	0.34
2	10,000	5	0.6	10,000	5.025	0.598	0.00	0.51	0.34
3	10,000	5	0.8	10,001	5.067	0.797	0.01	1.35	0.34

References

1. Drakonaki, E.E.; Allen, G.M.; Wilson, D.J. Ultrasound elastography for musculoskeletal applications. *Br. J. Radiol.* **2012**, *85*, 1435–1445. [CrossRef] [PubMed]
2. Sigrist, R.M.; Liau, J.; El Kaffas, A.; Chammas, M.C.; Willmann, J.K. Ultrasound elastography: Review of techniques and clinical applications. *Theranostics* **2017**, *7*, 1303. [CrossRef] [PubMed]
3. Zaleska-Dorobisz, A.U.; Kaczorowski, B.K.; Pawluś, B.A.; Puchalska, B.A.; Inglot, B.M. Ultrasound elastography–review of techniques and its clinical applications. *Brain* **2013**, *6*, 10–14. [CrossRef]
4. Ophir, J.; Cespedes, I.; Ponnekanti, H.; Yazdi, Y.; Li, X. Elastography: A quantitative method for imaging the elasticity of biological tissues. *Ultrason. Imaging* **1991**, *13*, 111–134. [CrossRef] [PubMed]
5. Cosgrove, D.; Piscaglia, F.; Bamber, J.; Bojunga, J.; Correas, J.-M.; Gilja, O.H.; Klauser, A.S.; Sporea, I.; Calliada, F.; Cantisani, V.; et al. EFSUMB Guidelines and Recommendations on the Clinical Use of Ultrasound Elastography.Part 2: Clinical Applications. *Ultraschall Med.* **2013**, *34*, 238–253. [CrossRef]
6. Barr, R.G.; Nakashima, K.; Amy, D.; Cosgrove, D.; Farrokh, A.; Schafer, F.; Bamber, J.C.; Castera, L.; Choi, B.I.; Chou, Y.-H.; et al. WFUMB Guidelines and Recommendations for Clinical Use of Ultrasound Elastography: Part 2: Breast. *Ultrasound Med. Biol.* **2015**, *41*, 1148–1160. [CrossRef] [PubMed]
7. Ferraioli, G.; Filice, C.; Castera, L.; Choi, B.I.; Sporea, I.; Wilson, S.R.; David Cosgrove, C.F.D.; Barr, R. WFUMB guidelines and recommendations for clinical use of ultrasound elastography: Part 3: Liver. *Ultrasound Med. Biol.* **2015**, *41*, 1161–1179. [CrossRef] [PubMed]
8. Barr, R.G.; Cosgrove, D.; Brock, M.; Cantisani, V.; Correas, J.M.; Postema, A.W.; Georg Salomon, M.T.; Dietrich, C.F. WFUMB guidelines and recommendations on the clinical use of ultrasound elastography: Part 5. Prostate. *Ultrasound Med. Biol.* **2017**, *43*, 27–48. [CrossRef] [PubMed]
9. Cosgrove, D.; Barr, R.; Bojunga, J.; Cantisani, V.; Chammas, M.C.; Dighe, M.; Vinayak, S.; Xu, J.-M.; Dietrich, C.F. WFUMB Guidelines and Recommendations on the Clinical Use of Ultrasound Elastography: Part 4. Thyroid. *Ultrasound Med. Biol.* **2017**, *43*, 4–26. [CrossRef] [PubMed]

10. Domenichini, R.; Pialat, J.-B.; Podda, A.; Aubry, S. Ultrasound elastography in tendon pathology: State of the art. *Skelet. Radiol.* **2017**, *46*, 1643–1655. [CrossRef] [PubMed]
11. Brandenburg, J.E.; Eby, S.F.; Song, P.; Zhao, H.; Brault, J.S.; Chen, S.; An, K.-N. Ultrasound Elastography: The New Frontier in Direct Measurement of Muscle Stiffness. *Arch. Phys. Med. Rehabil.* **2014**, *95*, 2207–2219. [CrossRef]
12. Lin, C.-Y.; Lin, C.-C.; Chou, Y.-C.; Chen, P.-Y.; Wang, C.-L. Heel Pad Stiffness in Plantar Heel Pain by Shear Wave Elastography. *Ultrasound Med. Biol.* **2015**, *41*, 2890–2898. [CrossRef] [PubMed]
13. Lin, C.-Y.; Chen, P.-Y.; Shau, Y.-W.; Tai, H.-C.; Wang, C.-L. Spatial-dependent mechanical properties of the heel pad by shear wave elastography. *J. Biomech.* **2017**, *53*, 191–195. [CrossRef]
14. Lin, C.-Y.; Wu, C.-H.; Özçakar, L. Restoration of Heel Pad Elasticity in Heel Pad Syndrome Evaluated by Shear Wave Elastography. *Am. J. Phys. Med. Rehabil.* **2017**, *96*, e96. [CrossRef] [PubMed]
15. Deng, C.X.; Hong, X.; Stegemann, J.P. Ultrasound Imaging Techniques for Spatiotemporal Characterization of Composition, Microstructure, and Mechanical Properties in Tissue Engineering. *Tissue Eng. Part B Rev.* **2016**, *22*, 311–321. [CrossRef] [PubMed]
16. Hong, X.; Stegemann, J.P.; Deng, C.X. Microscale characterization of the viscoelastic properties of hydrogel biomaterials using dual-mode ultrasound elastography. *Biomaterials* **2016**, *88*, 12–24. [CrossRef]
17. Hong, X.; Annamalai, R.T.; Kemerer, T.S.; Deng, C.X.; Stegemann, J.P. Multimode ultrasound viscoelastography for three-dimensional interrogation of microscale mechanical properties in heterogeneous biomaterials. *Biomaterials* **2018**, *178*, 11–22. [CrossRef] [PubMed]
18. Doherty, J.R.; Trahey, G.E.; Nightingale, K.R.; Palmeri, M.L. Acoustic radiation force elasticity imaging in diagnostic ultrasound. *IEEE Trans. Ultrason. Ferroelectr. Freq. Control* **2013**, *60*, 685–701. [CrossRef] [PubMed]
19. Palmeri, M.L.; Nightingale, K.R. Acoustic radiation force-based elasticity imaging methods. *Interface Focus* **2011**, *1*, 553–564. [CrossRef]
20. Kim, W.; Ferguson, V.L.; Borden, M.; Neu, C.P. Application of Elastography for the Noninvasive Assessment of Biomechanics in Engineered Biomaterials and Tissues. *Ann. Biomed. Eng.* **2016**, *44*, 705–724. [CrossRef] [PubMed]
21. Ryu, J.; Jeong, W.K. Current status of musculoskeletal application of shear wave elastography. *Ultrason.* **2017**, *36*, 185–197. [CrossRef]
22. Garteiser, P.; Doblas, S.; Daire, J.-L.; Wagner, M.; Leitao, H.; Vilgrain, V.; Sinkus, R.; Van Beers, B.E. MR elastography of liver tumours: Value of viscoelastic properties for tumour characterisation. *Eur. Radiol.* **2012**, *22*, 2169–2177. [CrossRef]
23. Qiu, Y.; Sridhar, M.; Tsou, J.K.; Lindfors, K.K.; Insana, M.F. Ultrasonic Viscoelasticity Imaging of Nonpalpable Breast Tumors: Preliminary Results. *Acad. Radiol.* **2008**, *15*, 1526–1533. [CrossRef] [PubMed]
24. Montagnon, E.; Tripette, J.; Mfoumou, E.; Cloutier, G. Acoustic radiation force induced elastography (ARFIRE): A new method to characterize blood clot viscoelastic properties. In Proceedings of the 2012 IEEE International Ultrasonics Symposium, Dresden, Germany, 7–10 October 2012; pp. 13–16.
25. Carrascal, C.A. Viscoelastic Creep Imaging. In *Ultrasound Elastography for Biomedical Applications and Medicine*; John Wiley & Sons: Hoboken, NJ, USA, 2018; pp. 171–188. [CrossRef]
26. Walker, W.F.; Fernandez, F.J.; Negron, L.A. A method of imaging viscoelastic parameters with acoustic radiation force. *Phys. Med. Biol.* **2000**, *45*, 1437–1447. [CrossRef] [PubMed]
27. Viola, F.; Walker, W.F. Radiation force imaging of viscoelastic properties with reduced artifacts. *IEEE Trans. Ultrason. Ferroelectr. Freq. Control.* **2003**, *50*, 736–742. [CrossRef]
28. Mauldin, F.; Haider, M.; Loboa, E.; Behler, R.; Euliss, L.; Pfeiler, T.; Gallippi, C. Monitored steady-state excitation and recovery (MSSR) radiation force imaging using viscoelastic models. *IEEE Trans. Ultrason. Ferroelectr. Freq. Control.* **2008**, *55*, 1597–1610. [CrossRef]
29. Amador, C.; Urban, M.W.; Chen, S.; Greenleaf, J.F. Loss tangent and complex modulus estimated by acoustic radiation force creep and shear wave dispersion. *Phys. Med. Biol.* **2012**, *57*, 1263. [CrossRef]
30. Amador, C.; Urban, M.W.; Chen, S.; Greenleaf, J.F. Complex shear modulus quantification from acoustic radiation force creep-recovery and shear wave propagation. In Proceedings of the 2012 IEEE International Ultrasonics Symposium, Dresden, Germany, 7–10 October 2012; pp. 1850–1853.
31. Amador, C.; Qiang, B.; Urban, M.W.; Chen, S.; Greenleaf, J.F. Acoustic radiation force creep-recovery: Theory and finite element modeling. In Proceedings of the 2013 IEEE International Ultrasonics Symposium (IUS), Prague, Czech Republic, 21–25 July 2013; pp. 363–366.
32. Selzo, M.R.; Gallippi, C.M. Viscoelastic response (VisR) imaging for assessment of viscoelasticity in voigt materials. *IEEE Trans. Ultrason. Ferroelectr. Freq. Control.* **2013**, *60*, 2488–2500. [CrossRef]
33. Nabavizadeh, A.; Kinnick, R.R.; Bayat, M.; Amador, C.; Urban, M.W.; Alizad, A.; Fatemi, M. Automated compression device for viscoelasticity imaging. *IEEE Trans. Biomed. Eng.* **2016**, *64*, 1535–1546. [CrossRef] [PubMed]
34. Bayat, M.; Nabavizadeh, A.; Kumar, V.; Gregory, A.; Insana, M.; Alizad, A.; Fatemi, M. Automated In Vivo Sub-Hertz Analysis of Viscoelasticity (SAVE) for Evaluation of Breast Lesions. *IEEE Trans. Biomed. Eng.* **2017**, *65*, 2237–2247. [CrossRef]
35. Lin, C.-Y. Alternative Form of Standard Linear Solid Model for Characterizing Stress Relaxation and Creep: Including a Novel Parameter for Quantifying the Ratio of Fluids to Solids of a Viscoelastic Solid. *Front. Mater.* **2020**, *7*, 11. [CrossRef]
36. Da Silva, R.J.B. Setting Target Measurement Uncertainty in Water Analysis. *Water* **2013**, *5*, 1279–1302. [CrossRef]

37. Baldewsing, R.A.; de Korte, C.L.; Schaar, J.A.; Mastik, F.; van der Steen, A.F. Finite element modeling and intravascular ultrasound elastography of vulnerable plaques: Parameter variation. *Ultrasonics* **2004**, *42*, 723–729. [CrossRef] [PubMed]
38. Caenen, A.; Shcherbakova, D.; Verhegghe, B.; Papadacci, C.; Pernot, M.; Segers, P.; Swillens, A. A versatile and experimentally validated finite element model to assess the accuracy of shear wave elastography in a bounded viscoelastic medium. *IEEE Trans. Ultrason. Ferroelectr. Freq. Control.* **2015**, *62*, 439–450. [CrossRef] [PubMed]
39. Maksuti, E.; Bini, F.; Fiorentini, S.; Blasi, G.; Urban, M.W.; Marinozzi, F.; Larsson, M. Influence of wall thickness and diameter on arterial shear wave elastography: A phantom and finite element study. *Phys. Med. Biol.* **2017**, *62*, 2694–2718. [CrossRef] [PubMed]
40. Pasyar, P.; Arabalibeik, H.; Mohammadi, M.; Rezazadeh, H.; Sadeghi, V.; Askari, M.; Mirbagheri, A. Ultrasound elastography using shear wave interference patterns: A finite element study of affecting factors. *Phys. Eng. Sci. Med.* **2021**, *44*, 253–263. [CrossRef] [PubMed]
41. Hollis, L.; Barnhill, E.; Conlisk, N.; Thomas-Seale, L.E.; Roberts, N.; Pankaj, P.; Hoskins, P.R. Finite element analysis to compare the accuracy of the direct and mdev inversion algorithms in MR elastography. *IAENG Int. J. Comput. Sci.* **2016**, *43*, 137–146.
42. Hollis, L.; Thomas-Seale, L.; Conlisk, N.; Roberts, N.; Pankaj, P.; Hoskins, P.R. Investigation of modelling parameters for finite element analysis of MR elastography. In *Computational Biomechanics for Medicine*; Springer: Cham, Switzerland, 2016; pp. 75–84.
43. Thomas-Seale LE, J.; Hollis, L.; Klatt, D.; Sack, I.; Roberts, N.; Pankaj, P.; Hoskins, P.R. The simulation of magnetic resonance elastography through atherosclerosis. *J. Biomech.* **2016**, *49*, 1781–1788. [CrossRef]
44. Hollis, L.; Barnhill, E.; Perrins, M.; Kennedy, P.; Conlisk, N.; Brown, C.; Hoskins, P.R.; Pankaj, P.; Roberts, N. Finite element analysis to investigate variability of MR elastography in the human thigh. *Magn. Reson. Imaging* **2017**, *43*, 27–36. [CrossRef]
45. Lin, C.Y.; Lin, S.R. Investigating the accuracy of ultrasound viscoelastic creep imaging for measuring the viscoelastic properties of a single-inclusion phantom. *Int. J. Mech. Sci.* **2021**, *199*, 106409. [CrossRef]

 *materials*

Article

Topology Optimization-Based Damage Identification Using Visualized Ultrasonic Wave Propagation

Kazuki Ryuzono [1], **Shigeki Yashiro** [1,*], **Hiroto Nagai** [1] and **Nobuyuki Toyama** [2]

[1]   Department of Aeronautics and Astronautics, Kyushu University, 744 Motooka, Nishi-ku, Fukuoka 819-0395, Japan; Ryuzono@aero.kyushu-u.ac.jp (K.R.); nagai@aero.kyushu-u.ac.jp (H.N.)

[2]   National Metrology Institute of Japan, National Institute of Advanced Industrial Science and Technology (AIST), 1-1-1 Umezono, Tsukuba, Ibaraki 305-8568, Japan; toyama-n@aist.go.jp

*   Correspondence: yashiro@aero.kyushu-u.ac.jp; Tel.: +81-92-802-3030

Received: 11 November 2019; Accepted: 18 December 2019; Published: 19 December 2019

Abstract: This study proposes a new damage identification method based on topology optimization, combined with visualized ultrasonic wave propagation. Although a moving diagram of traveling waves aids in damage detection, it is difficult to acquire quantitative information about the damage, for which topology optimization is suitable. In this approach, a damage parameter, varying Young's modulus, represents the state of the damage in a finite element model. The feature of ultrasonic wave propagation (e.g., the maximum amplitude map in this study) is inversely reproduced in the model by optimizing the distribution of the damage parameters. The actual state of the damage was successfully estimated with high accuracy in numerical examples. The sensitivity of the objective function, as well as the appropriate penalization exponent for Young's modulus, was discussed. Moreover, the proposed method was applied to experimentally measured wave propagation in an aluminum plate with an artificial crack, and the estimated damage state and the sensitivity of the objective function had the same tendency as the numerical example. These results demonstrate the feasibility of the proposed method.

Keywords: non-destructive inspection; damage identification; topology optimization; ultrasonic wave propagation; ultrasonic visualization

---

1. Introduction

Quantitative non-destructive inspection is important to ensure the reliability and safety of structures such as aircraft and automobiles [1]. Among the non-destructive inspection techniques, ultrasonic inspection has been widely used because ultrasonic waves are highly sensitive to a damaged part and propagate over long distances. The importance of quantitative ultrasonic inspection has been described for more than 30 years [2].

To undertake quantitative evaluation by ultrasonic inspection, it is essential to develop both the measurement techniques and data analyses. In general, ultrasonic signals contain reflected waves, diffracted waves, and mode-converted waves, and some kinds of ultrasonic waves such as Lamb waves and Love waves have dispersive nature [3–5], which makes ultrasonic waveforms challenging to interpret. Consequently, evaluation of the damage is dependent on the skills of engineers, making misreading signals and false recognition of defects inevitable. Furthermore, the conventional ultrasonic inspection process is not automated; thus, the inspections require a lot of time and labor to scan the whole structure. Numerous studies [4,6–13] have been reported about new techniques and data analyses to overcome these difficulties. For example, ultrasonic arrays [6] have improved the inspection quality and have reduced the inspection costs by performing beam steering with a wide viewing angle through controlled transmission of multiple elements. Acoustic emission detection using a fiber-optic sensor and mode analysis [7–9] has also been developed to achieve a quantitative evaluation of the damage in composite laminates. The wavelet-transform has frequently been incorporated with the signal processing to analyze dispersive waves [4,10,11]. Furthermore, some inverse analyses [4,12,13] have been presented to estimate the damage quantitatively using the artificial neural network and genetic algorithm.

Aiming at further damage visibility and operability, a visualization method of ultrasonic wave propagation [14–29] has been developed. In this method, ultrasonic waves are generated by illuminating a specimen surface with a pulsed laser and are received by a fixed transducer [14]. Based on the reciprocity of wave propagation [15], the amplitude of each waveform at a particular time plotted in a contour map yields a moving diagram of the wave propagation from the receiver. The feasibility of this method was demonstrated by applying it to a crack and an artificial hollow in metallic materials [15], delamination in carbon fiber reinforced plastic (CFRP) laminates [16–20], a crack in welded steel plates [21], and disbonds in adhesively bonded CFRP/aluminum joints [22]. Frequency and/or wavenumber domain analysis using the Fourier- or wavelet-transform was introduced into ultrasonic propagation imaging to easily interpret the visualized results of wave propagation by isolating a specific frequency mode [23–25]. Moreover, a fully non-contact ultrasonic inspection [26–28] was demonstrated by replacing a fixed transducer with a laser Doppler vibrometer, and this method removed ringing due to the resonance of the piezoelectric transducer and made it easy to interpret scattered waves [28].

Although the visualization method of ultrasonic propagation has high damage visibility and excellent operability, it is difficult to evaluate the damage quantitatively. An efficient automatic ultrasonic image analysis has been presented using deep learning [29], but at the moment, the main target of this method is automated damage detection, not quantitative evaluation. A moving diagram of wave propagation includes wave signals at all illuminating points. Therefore, appropriate analysis for all wave signals will have the potential to acquire quantitative information about the damage.

Topology optimization [30] will be suitable for that purpose, but to our knowledge, there are few studies that apply it to damage identification. In topology optimization for structural design [31–33], a design domain is discretized by finite elements, and the material density distribution is assigned as design variables. Similarly, in damage identification problems, the damage severity is the design variable instead of the material density, and the damage distribution is inversely estimated by reproducing an input phenomenon of focus. Based on this idea, Lee et al. [34] demonstrated numerical examples for estimating damage in thin plates and beam models, taking resonant and anti-resonant frequencies as an objective function. Some numerical examples that focused on natural frequencies were also reported [35–37]. Niemann et al. [38–40] estimated the approximate location of the damage in CFRP laminates after impact tests. However, this damage identification focusing on frequency characteristics was not very accurate. The reason for this is the fact that frequency characteristics are not sufficiently sensitive to damage.

This study proposes a damage identification method using the visualization technique of ultrasonic wave propagation. To this end, we incorporate topology optimization with a moving diagram of wave propagation, having high sensitivity to damage. The feature of wave propagation is reproduced in the analytical model by optimizing the distribution of the damage parameters. As a result, quantitative information about the damage is estimated. This study is the first attempt to integrate the inverse analysis based on topology optimization with the ultrasonic imaging inspection. The method is first proposed, and its feasibility is then verified using a two-dimensional case of known damage location and size.

2. Damage Identification Procedure

2.1. Concept of Damage Identification

Figure 1 presents the sequential process flow of the proposed damage identification method. The proposed method estimates the damage distribution that reproduces a moving diagram of ultrasonic wave propagation through the following steps. (1) Ultrasonic imaging inspection is conducted, and an ultrasonic feature in the moving diagram is adopted as target data. A potential damaged site is extracted from an extensive inspection area as the design domain. (2) An analysis model of the extracted domain is discretized into finite elements, and the initial damage parameter is assigned to each element. (3) Finite element analysis of ultrasonic propagation is performed, and the ultrasonic feature is evaluated at the present damage distribution. (4) The error between the estimated ultrasonic feature and the target data is calculated. Then, (5) the correct number of damage parameters is explored by mathematical programming. (6) The damage parameters are updated, and the process is repeated from step (3) until convergence. As a result, the position, size, and shape of damage are obtained on a per-element basis.

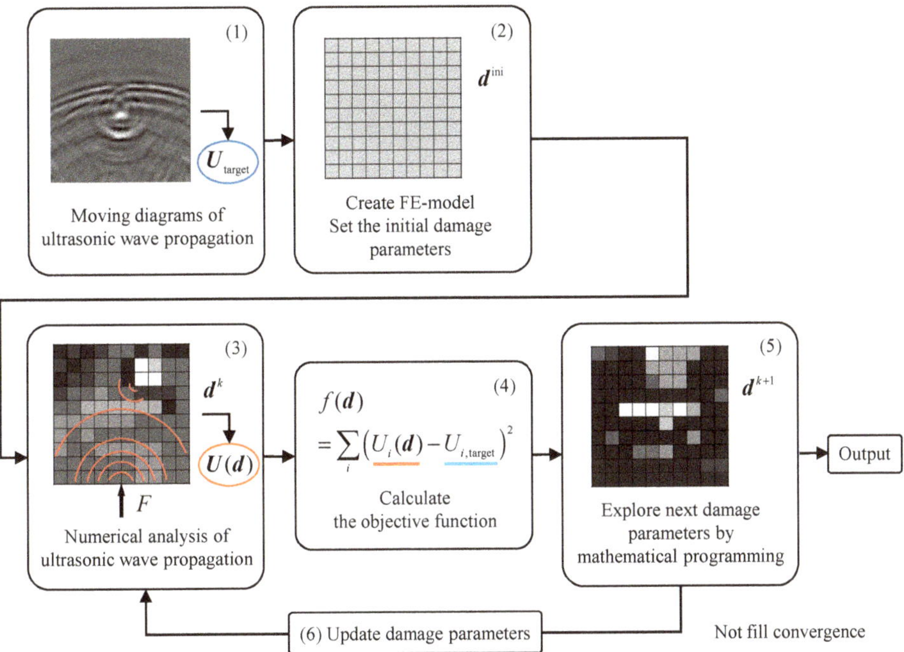

Figure 1. The conceptual flowchart of damage identification based on topology optimization.

2.2. Setup of Optimization Problem

The design domain D is discretized into finite elements, and the damage parameter d_i is assigned to element i:

$$d_i = \begin{cases} 0 & \text{for } i \in \Omega_d \quad \text{damage} \\ 1 & \text{for } i \in D\backslash\Omega_d \quad \text{no damage} \end{cases} \tag{1}$$

where Ω_d is the damaged domain, and $D\backslash\Omega_d$ is the intact domain. The solid isotropic material with penalization (SIMP) method [31] is used to relax the damage parameter, and a continuous value from 0 to 1 represents the severity of the damage. With this setting, the damage identification problem is translated into a distribution problem of the damage parameter in the design domain D.

The Young's modulus of element i is expressed as a function of the damage parameter d_i as:

$$E_i(d_i) = (E_1 - E_0)d_i^p + E_0, \tag{2}$$

where E_1 and E_0 are Young's modulus of a solid (i.e., perfectly intact) and in a void (i.e., perfectly damaged), and p is the penalization exponent for intermediate damage. Although E_0 is initially zero, elimination of an element (i.e., modification of the model) is cumbersome in the optimization process; therefore, a small stiffness ($E_0 = 0.001$ MPa) is assigned to perfectly damaged elements.

The topology optimization problem is defined as the minimization of the square error between the estimated ultrasonic feature and the target data as follows:

$$\min_{d} f(d) = \sum_{j=1}^{m} \left(U_j(d) - U_{j,\text{target}} \right)^2 \quad \text{subject to} \quad 0 \le d_i \le 1 \quad \text{for } i = 1, \cdots, n \tag{3}$$

where $U_j(d)$ is the analytically obtained ultrasonic feature at node j with the present damage state d, $U_{j,\text{target}}$ is the ultrasonic feature of target data at an illuminating point, and m and n are the total number of nodes and elements in the design domain D. The ultrasonic feature in the objective function should be selected appropriately, depending on the problem. Although a few studies [38–40] used the summation of damage parameters as a constraint condition, no constraint conditions other than Equation (3) provided better results in the preliminary investigation.

3. Numerical Examples

To verify the feasibility of the proposed method, this section applies it to numerical examples with known damage. The target data ultrasonic feature $U_{j,\text{target}}$ in Equation (3) is obtained by forward analysis of ultrasonic wave propagation in a target model. To simplify the problem, a penetrating crack in a metal plate is considered here, and all the analyses are two-dimensional.

3.1. Numerical Models

Figure 2a shows the analysis model. The target plate model was 100×100 mm, and a penetrating crack 6 mm long and 1 mm wide was introduced at the center. Four-node quadrilateral solid (plane-stress) elements were used, and the size of each element was 1×1 mm. There were 10,000 elements and 10,201 nodes. The crack was expressed as damaged elements with a small Young's modulus E_0. Five period-sinusoidal wave loads multiplied by the Hanning window function were applied to all nodes on the lower surface; their amplitude and frequency were 1 N and 1 MHz. A non-reflective boundary condition [41] that removes one reflection was set on the upper, left, and right surfaces. The material was assumed to be stainless steel, and its Young's modulus, Poisson's ratio, and density were 186.6 GPa, 0.306, and 7.86 g/cm^3, respectively.

Figure 3 depicts the analyzed ultrasonic wave propagation. The ultrasounds propagated from the bottom to the top and were diffracted and scattered at the crack. The waves propagating from the bottom corners were reflected waves. Figure 3b shows the distribution of the maximum amplitude of the mean stress. The maximum amplitude had a characteristic distribution near the crack (e.g., high in the lower area and low in the upper area to the actual crack). Therefore, the crack will be reproduced by focusing on the maximum amplitude distribution.

The inverse analysis model (i.e., design domain D) was a 10×10 mm domain (including 100 elements and 121 nodes), which was extracted from the center of the target model, as illustrated in Figure 2b. Topology optimization defines the same number of design variables as elements in the design domain. Therefore, the size of the inverse analysis model and the calculation cost are in a trade-off relationship. Input loads, similar to those in the forward analysis, were applied to the lower surface of the inverse analysis model. A non-reflective boundary condition [41] was set on the upper, left, and right surfaces.

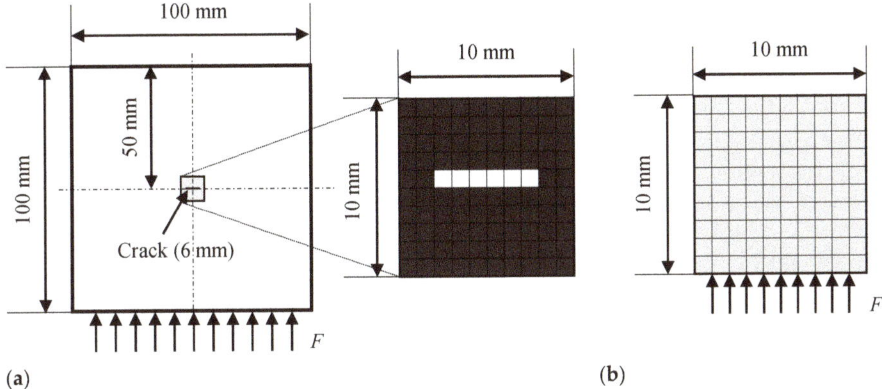

Figure 2. Models to calculate data of ultrasonic waves: (**a**) Target model; (**b**) Inverse analysis model.

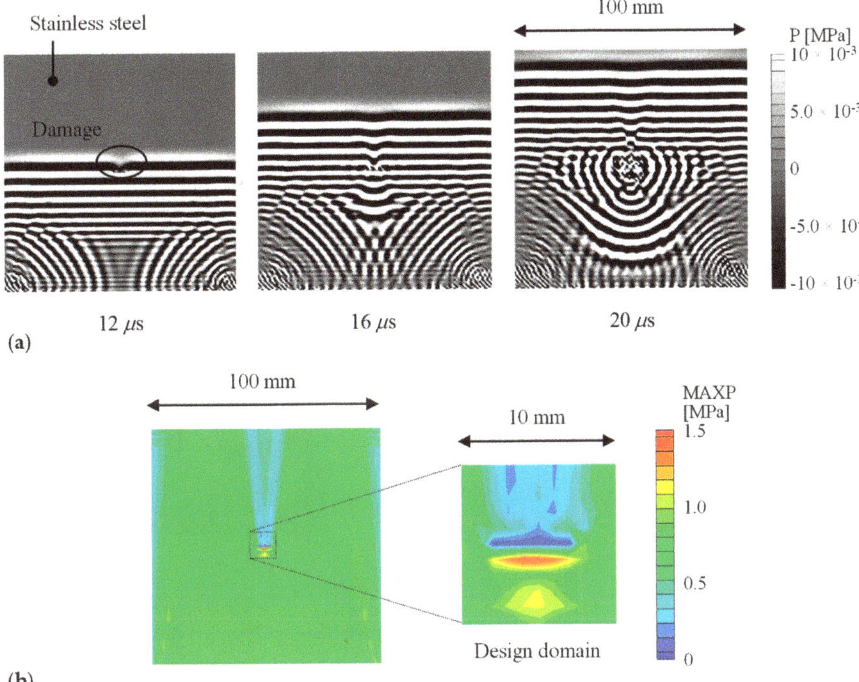

Figure 3. Numerical results of wave propagation on a stainless-steel plate with a crack: (**a**) Snapshots (mean stress); (**b**) Distribution of the maximum amplitude of the mean stress.

3.2. Damage Identification for Numerical Results

The maximum amplitude of the mean stress was adopted as the ultrasonic feature in the objective function. Thus, the optimization problem is defined as follows:

$$\min_{d} f(\boldsymbol{d}) = \sum_{j=1}^{m} \left(\sigma_{j,\max}(\boldsymbol{d}) - \sigma_{j,\max,\text{target}} \right)^2 \quad \text{subject to} \quad 0 \le d_i \le 1 \quad \text{for } i = 1, \cdots, n \tag{4}$$

where $\sigma_{j,\max}(\boldsymbol{d})$ is the maximum amplitude of the mean stress estimated at node j with the present damage state $\boldsymbol{d}$, and $\sigma_{j,\max,\text{target}}$ is that of the target data at node j obtained by forward analysis (Figure 3b right) in this section. A constant value of the damage parameter, d^{ini}, was initially assigned to all the elements, and ten inverse analyses were performed using d^{ini} from 0.1 to 1.0, with an increment of 0.1. The penalization exponent p of 3 was used as in previous studies [31,34–36,38]. Sequential quadratic programming in the MATLAB Optimization Toolbox (R2019a, MathWorks, Inc, Natick, MA, USA) was used for exploration and updating of the damage parameters. The gradient of the objective function was calculated by the finite-difference method.

Figure 4a depicts typical damage distributions estimated by the proposed method. The estimated damage parameters of the elements in the actual damaged domain (T) were smaller than that of most elements in the actual intact domain ($D\backslash T$) in all of the ten cases of the initial damage parameter d^{ini} distribution. This result suggests that the damage in the target can be estimated by the proposed method; however, the estimated damage states were different depending on the initial damage parameter d^{ini} in the upper area of $D\backslash T$. Moreover, for example, in the case of $d^{\text{ini}} = 0.5$, the damage parameters smaller than its maximum in T (0.2016) were estimated in nine elements in $D\backslash T$, which were erroneously determined to be damaged. Such elements were concentrated in the upper region of D.

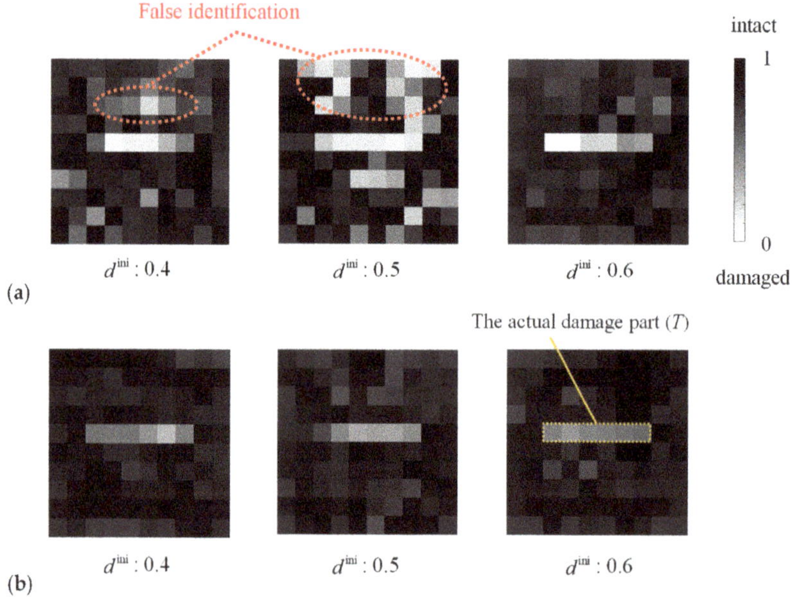

Figure 4. The optimal solution of the damage state obtained using ultrasonic wave data (**a**) from a single direction and (**b**) from two directions.

The reason for this false identification is caused by the low sensitivity of the objective function in the area above the crack. The objective function was evaluated at various damage states depicted in Figure 5a. Damaged elements were added to the lower area of $D\backslash T$ in #1–#5 and to the upper area in #7–#10. Figure 5b presents the value of the objective function obtained by forward analysis in these damage states. The gradient of the objective function within #6–#10 was smaller than that within #1–#6. Moreover, the maximum amplitude in the area above the actual crack was smaller than that in the other areas, as shown in Figure 3b, because the diffracted ultrasonic waves propagated with a small amplitude above the crack as a result of the high directivity of ultrasound. Therefore, the maximum amplitude hardly changed in the area above the actual crack even if the damage state in that area was altered. Consequently, the damage identification results depended on the relationship between the crack geometry and the direction of ultrasonic wave propagation, and it was difficult to estimate the right damage state in the area above the crack with the objective function (4).

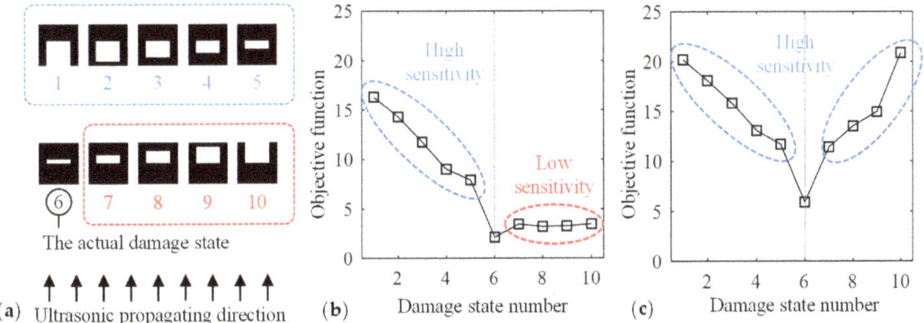

Figure 5. The objective function values obtained using ultrasonic wave data in (**a**) the various damage states (**b**) from a single direction and (**c**) from two directions.

To improve the sensitivity of the objective function, the wave propagation from the top to bottom (superscript: upper) was considered in addition to that from the bottom to the top (superscript: lower); thus the optimization problem is re-defined as follows:

$$\min_{d} f(\boldsymbol{d}) = \sum_{j=1}^{m} \left(\sigma_{j,\max}^{\text{lower}}(\boldsymbol{d}) - \sigma_{j,\max,\text{target}}^{\text{lower}} \right)^2 + \sum_{j=1}^{m} \left(\sigma_{j,\max}^{\text{upper}}(\boldsymbol{d}) - \sigma_{j,\max,\text{target}}^{\text{upper}} \right)^2$$

$$\text{subject to} \quad 0 \leq d_i \leq 1 \quad \text{for } i = 1, \cdots, n$$

(5)

In ultrasonic visualization experiments, ultrasonic propagation data from two directions can easily be recorded.

Figure 4b depicts the estimated damage states based on Equation (5). The damage parameter in $D \backslash T$ was always greater than its maximum in T, and thus, the actual damage T was successfully identified. However, the difference of the damage parameter in T and in $D \backslash T$ was small, and this issue will be discussed in the following section. The sensitivity of the objective function was high in both the upper and lower areas of the target crack, as shown in Figure 5c. Thus, when the sensitivity was enhanced in the entire design domain, the proposed method provided the damage identification results closer to the target damage state.

The above results and discussion demonstrate the feasibility of the proposed method. This study used the maximum amplitude distribution as the ultrasonic feature, and two sets of ultrasonic propagation data were required in the objective function to enhance its sensitivity. However, considering the cost of inspection, it is recommended to estimate damage using a single set of wave propagation data. To that end, the objective function should be improved and will be investigated in our future study.

3.3. Effect of the Penalization Exponent

The penalization exponent p in Equation (2) has been discussed in previous studies, because the relaxation of the design variables yields an intermediate density called a gray-scale element, which cannot be interpreted physically in structural design problems. To clarify the physical meaning of the gray-scale element, Bendsøe et al. [42] discussed the range of p based on Hashin-Shtrikman (HS) bounds [43] and proved that the SIMP method is physically permissible as long as p is greater than a certain value (e.g., $p \geq 3$ for two-dimensional problems with Poisson's ratio of 1/3). In previous studies of damage identification based on topology optimization, Nishizu and Neumann et al. [35,38–40] set $p = 3$, Reumers et al. [36] set $p = 1$, and Eslami et al. [37] changed p gradually from 3 to 1. An element (i.e., microscopic area) will have various Young's modulus depending on the severity of the damage. Therefore, unlike structural design problems, the intermediate damage parameters are acceptable in the damage identification problem. Instead, it is essential to set a threshold to distinguish a damaged from an intact region.

Damage identification using various penalty exponents and the HS upper bound was performed. Here, the problem is defined by Equation (5) and Figure 2; the initial value of the damage parameter, d^{ini}, was 0.5 in all the cases. Figure 6a depicts the estimated damage states, and Figure 6b shows a variation of Young's modulus normalized by E_1. The inverse analysis did not converge when p was less than 0.5 or greater than 3. In all converged results, the maximum value of the damage parameter in T was smaller than its minimum in $D\backslash T$, and the damage state was estimated appropriately.

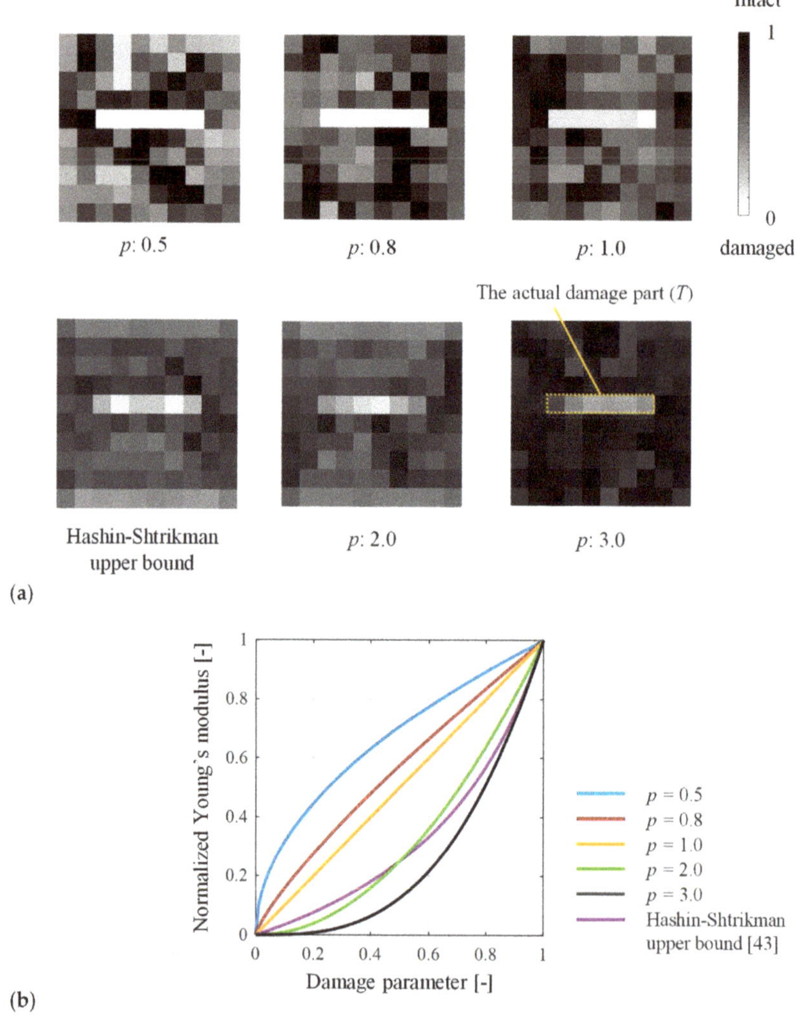

Figure 6. Comparison of (a) the optimal solution of the damage state and (b) the values of Young's modulus as a function of damage parameters with the various penalization exponents and the Hashin-Shtrikman upper bound [43].

When p was greater than 1 (including the HS upper bound), the damage parameters were almost uniform in $D\backslash T$. Furthermore, as p increased, the damage parameters in $D\backslash T$ approached 1, and the number of gray-scale elements decreased. This is because the gradient of Young's modulus in damage parameters close to 1 is large when $p = 2$ or more, as shown in Figure 6b. Therefore, p should be greater than 2 to interpret the elements as undamaged physically.

On the other hand, when p was 1 or less, the difference of the damage parameter in T and in $D\backslash T$ was larger than that with p greater than 1. Furthermore, the damage parameter in T became almost 0. As shown in Figure 6b, when the damage parameter is close to 0, Young's modulus changes significantly with a small variation in the damage parameter. Therefore, the damage parameter tends towards 0 in T, and it hardly approaches 0 in $D\backslash T$. The penalization exponent of 0.5–1 enables setting a threshold to distinguish between damaged elements from intact ones. For example, when using $p = 0.8$ and the threshold $d_{\text{th}} = 0.1$, the damaged and undamaged elements were distinguished in all initial values of the damage parameter. The above discussion indicates that the penalization exponents p of 0.5–1 are suitable for damage identification.

4. Application

In this section, the feasibility of the proposed method is verified by applying it to the measured wave propagation in an aluminum plate with an artificially generated crack whose position, length, and width are known. Therefore, the experimentally measured maximum amplitude was used as the target data $\sigma_{j,\text{max,target}}$ in Equation (4).

4.1. Experiment

Figure 7 presents a schematic of the test configuration. The test piece was an aluminum plate with dimensions of $500 \times 500 \times 5$ mm. A crack 4 mm long and less than 0.3 mm wide was introduced at the center of it. A 40×40 mm region, including the crack, was scanned by an Nd:YAG pulsed laser (DIVA II, Thales Laser Co., Ltd) at a constant illumination spacing of 0.25 mm (160×160 points). The pulse duration was 10 ns with 5 mJ energy per pulse, and the diameter of the beam spot was 2 mm. The ultrasonic wave generated by each illumination was received by a transducer placed 40 mm apart from the crack. Its resonance frequency was 5 MHz, and a 70° wedge was used. The received signals were recorded in a computer through an amplifier and a digital oscilloscope.

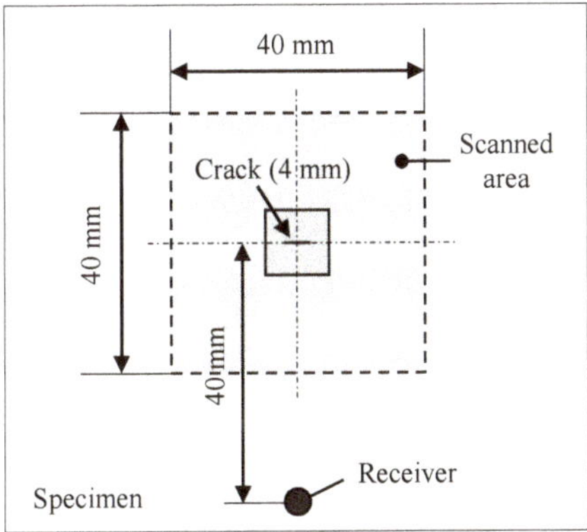

Figure 7. Schematic of an aluminum plate with a crack and the pulsed laser scanning area.

Figure 8a depicts the visualized results of the ultrasonic wave propagation in an aluminum plate with a crack. The traveling waves were reflected at the crack, and the amplitude below the crack was high due to the interference of the traveling and reflected waves, as shown in the maximum amplitude map (Figure 8b). In contrast, the diffracted waves propagated with a low amplitude because of the high directivity of ultrasound, hence the maximum amplitude above the crack was lesser than that in other regions. The approximate location of damage was found from the sudden amplitude change; however, it was difficult to quantitatively evaluate the length because the maximum amplitude distribution was rounded off near the crack tips, as shown in Figure 8b.

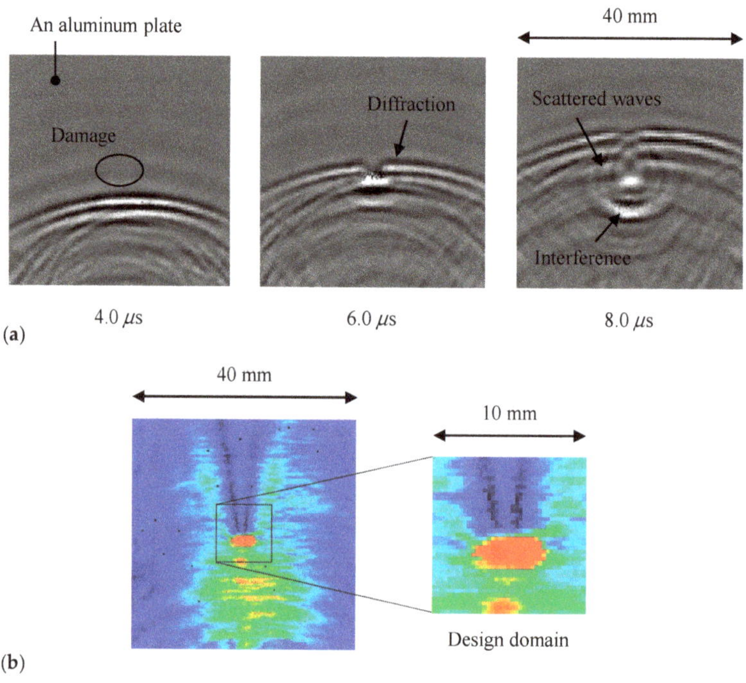

Figure 8. Inspection results by the visualization method of wave propagation on an aluminum plate with a crack: (**a**) Visualized results of ultrasonic waves; (**b**) Distribution of the maximum amplitude.

4.2. Damage Identification for Experimental Results

Figure 9a shows the two-dimensional analysis model. The dimension of the analysis domain was 50 mm long and 50 mm wide. Four-node quadrilateral solid (plane-stress) elements were used, and the size of each element was 1×1 mm. There were 2500 elements and 2601 nodes. Five period-sinusoidal wave loads multiplied by the Hanning window function were applied to the center point of the lower surface (i.e., the position of the receiver); their amplitude and frequency were 1 N and 5 MHz. A non-reflective boundary condition [41] that removes one reflection was set on the upper, left, and right surfaces. Young's modulus, Poisson's ratio, and density of the aluminum plate were 69.0 GPa, 0.34, and 2.70 g/cm^3.

The design domain D was a part of the analytical model located 40 mm from the bottom, as shown in Figure 9b. Its size was 10×10 mm, including 100 elements. The damage parameters only in the design domain D were updated, and those in the other areas remained constant at unity. The size of the element was 1 mm square, and therefore, the location and length of the crack could be identified, but the width could not be estimated. Equation (4) was used as the optimization problem setting, and the maximum amplitude distribution on the right of Figure 8b was used as the target data. The initial value of the damage parameter d^{ini} was 0.5. The penalization exponent $p = 0.8$ was adopted based on the discussion in Section 3.3.

Figure 10a depicts the estimated damage state. The maximum value of the damage parameter in T was 0.2950, and there were five elements in $D \backslash T$ that had a damage parameter smaller than that value. Most of these elements were in the area above the actual crack, and this pattern was similar to Figure 4a. The sensitivity of the objective function above the damage was evaluated in Figure 10b in the same manner as in Figure 5. The curve had the same tendency as in Figure 5b, and the low sensitivity in the area above the actual crack caused the misestimation. Nonetheless, as shown in Figure 10c, when using the threshold $d_{th} = 0.1$ for distinguishing between the damaged from the intact region, three elements out of the actual damage region T (four elements) were estimated as damaged, and all the other elements were estimated as intact.

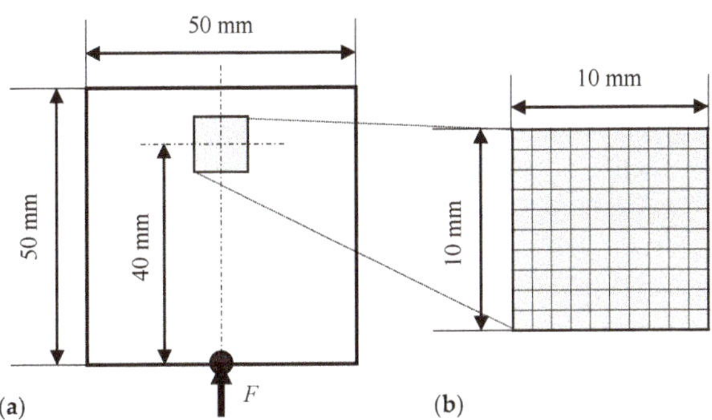

Figure 9. Numerical models with the same configuration as the experiment: (**a**) Ultrasonic wave propagation analysis model; (**b**) Inverse analysis model.

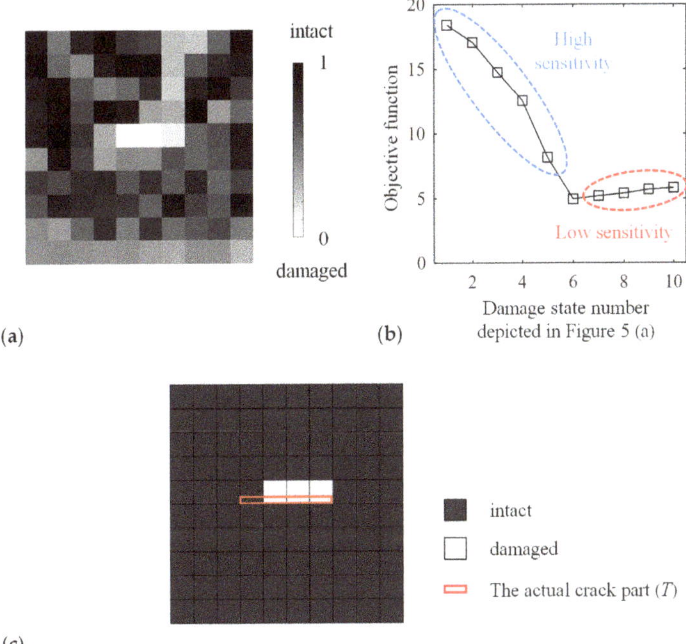

Figure 10. Results of applying the proposed method to experimentally measured wave propagation: (**a**) The optimal solution of the damage state; (**b**) The objective function values in the various damage states as with Figure 5; (**c**) The estimated damage state when using the threshold $d_{th} = 0.1$.

The feasibility of the proposed method was thus demonstrated, though these results were obtained in simple and ideal model cases. A more detailed investigation will be required for the application of the present method to real problems.

5. Conclusions

This study proposed a damage identification method of incorporating topology optimization with the visualization of ultrasonic wave propagation. The distribution of the damage parameter in a design domain was estimated by reproducing the ultrasonic maximum amplitude map of the target data. The proposed method was verified by applying it to numerical and experimental examples with known damage. The conclusions are summarized below:

1. The actual damage state was successfully estimated in the numerical examples. The damage was identified with high accuracy using ultrasonic propagation data from two directions to increase the sensitivity of the objective function in the whole design domain.
2. The effect of the penalization exponent p for determining Young's modulus and the efficiency of the damage identification was investigated. The standard value ($p = 3$) in the structural design problems resulted in moderate damage severity in the actual damaged area. In contrast, a penalization exponent less than unity resulted in higher damage severity in that area, and the difference between the damage parameter value between 'damaged' and 'intact' regions was greater than that in the case of $p = 3$. The penalization exponent within the range $0.5 \leq p \leq 1$ enabled setting a threshold to distinguish between damaged elements from intact ones.
3. The proposed method was applied to experimentally measured wave propagation in an aluminum plate with an artificial crack. The actual damaged area was estimated, and the estimated damage state and the sensitivity of the objective function had the same tendency as a similar numerical example. This demonstrated the feasibility of the proposed method.

Damage identification in simple and ideal model cases was demonstrated as the first attempt. In our future research, potential of the proposed method will be explored in general cases such as oblique incident to an embedded crack.

Author Contributions: Conceptualization, K.R. and S.Y.; methodology, K.R., S.Y., and H.N.; software, K.R.; validation, K.R., S.Y., and N.T.; formal analysis, K.R.; investigation, K.R., S.Y., and N.T.; resources, S.Y.; data curation, K.R.; writing—original draft preparation, K.R.; writing—review and editing, S.Y., H.N., and N.T.; visualization, K.R.; supervision, S.Y.; project administration, S.Y. All authors have read and agreed to the published version of the manuscript.

Funding: This research received no external funding.

Conflicts of Interest: The authors declare no conflict of interest.

References

1. Achenbach, J.D. Quantitative nondestructive evaluation. *Int. J. Solids Struct.* **2000**, *37*, 13–27. [CrossRef]
2. Thompson, R.B. Quantitative ultrasonic nondestructive evaluation methods. *J. Appl. Mech.* **1983**, *50*, 1191–1201. [CrossRef]
3. Rose, J.L. *Ultrasonic Waves in Solid Media*; Cambridge University Press: Cambridge, UK, 1999; pp. 101–142.
4. Su, Z.; Ye, L.; Lu, Y. Guided Lamb waves for identification of damage in composite structures: A review. *J. Sound Vib.* **2006**, *295*, 753–780. [CrossRef]
5. Moilanen, P. Ultrasonic guided waves in bone. *IEEE T. Ultrason. Ferr.* **2008**, *55*, 1277–1286. [CrossRef] [PubMed]
6. Drinkwater, B.W.; Wilcox, P.D. Ultrasonic arrays for non-destructive evaluation: A review. *NDT & E Int.* **2006**, *39*, 525–541.
7. Takeda, S.; Okabe, Y.; Takeda, N. Delamination detection in CFRP laminates with embedded small-diameter fiber Bragg grating sensors. *Compos. Part A.* **2002**, *33*, 971–980. [CrossRef]

8. Okabe, Y.; Fujibayashi, K.; Shimazaki, M.; Soejima, H.; Ogisu, T. Delamination detection in composite laminates using dispersion change based on mode conversion of Lamb waves. *Smart Mater. Struct.* **2010**, *19*, 115013. [CrossRef]

9. Yu, F.; Okabe, Y.; Wu, Q.; Shigeta, N. A novel method of identifying damage types in carbon fiber-reinforced plastic cross-ply laminates based on acoustic emission detection using a fiber-optic sensor. *Compos. Sci. Technol.* **2016**, *135*, 116–122. [CrossRef]

10. Veroy, K.L.; Wooh, S.-C.; Shi, Y. Analysis of dispersive waves using the wavelet transform. In *Review of Progress in Quantitative Nondestructive Evaluation*; Thompson, D.O., Chimenti, D.E., Eds.; Springer: Boston, MA, USA, 1999; Volume 18 A, pp. 687–694.

11. Choi, J.; Hong, J.-W. Characterization of wavelet coefficients for ultrasonic signals. *J. Appl. Phys.* **2010**, *107*, 114909. [CrossRef]

12. Barry, T.J.; Kesharaju, M.; Nagarajah, C.R.; Palanisamy, S. Defect characterisation in laminar composite structures using ultrasonic techniques and artificial neural networks. *J. Compos. Mater.* **2015**, *50*, 861–871. [CrossRef]

13. Oishi, A.; Yamada, K.; Yoshimura, S.; Yagawa, G. Quantitative nondestructive evaluation with ultrasonic method using neural networks and computational mechanics. *Comput. Mech.* **1995**, *15*, 521–533. [CrossRef]

14. Takatsubo, J.; Wang, B.; Tsuda, H.; Toyama, N. Generation laser scanning method for the visualization of ultrasounds propagating on a 3-D object with an arbitrary shape. *J. Solid Mech. Mater. Eng.* **2007**, *1*, 1405–1411. [CrossRef]

15. Yashiro, S.; Takatsubo, J.; Miyauchi, H.; Toyama, N. A novel technique for visualizing ultrasonic waves in general solid media by pulsed laser scan. *NDT E Int.* **2008**, *41*, 137–144.

16. Yashiro, S.; Takatsubo, J.; Toyama, N. An NDT technique for composite structures using visualized Lamb-wave propagation. *Compos. Sci. Technol.* **2007**, *67*, 3202–3208. [CrossRef]

17. Sohn, H.; Dutta, D.; Yang, J.Y.; Park, H.J.; DeSimio, M.; Olson, S.; Swenson, E. Delamination detection in composites through guided wave field image processing. *Compos. Sci. Technol.* **2011**, *71*, 1250–1256. [CrossRef]

18. Chia, C.C.; Lee, J.-R.; Park, C.-Y.; Jeong, H.-M. Laser ultrasonic anomalous wave propagation imaging method with adjacent wave subtraction: Application to actual damages in composite wing. *Opt. Laser Technol.* **2012**, *44*, 428–440. [CrossRef]

19. Lee, J.-R.; Chia, C.C.; Park, C.-Y.; Jeong, H. Laser ultrasonic anomalous wave propagation imaging method with adjacent wave subtraction: Algorithm. *Opt. Laser Technol.* **2012**, *44*, 1507–1515. [CrossRef]

20. Kudela, P.; Radzienski, M.; Ostachowicz, W. Impact induced damage assessment by means of Lamb wave image processing. *Mech. Syst. Signal Pr.* **2018**, *102*, 23–36. [CrossRef]

21. Yashiro, S.; Toyama, N.; Takatsubo, J.; Shiraishi, T. Laser-generation based imaging of ultrasonic wave propagation on welded steel plates and its application to defect detection. *Mater. Trans.* **2010**, *51*, 2069–2075. [CrossRef]

22. Toyama, N.; Yamamoto, T.; Urabe, K.; Tsuda, H. Ultrasonic inspection of adhesively bonded CFRP/aluminum joints using pulsed laser scanning. *Adv. Compos. Mater.* **2019**, *28*, 27–35. [CrossRef]

23. Lee, J.-R.; Chia, C.C.; Shin, H.J.; Park, C.-Y.; Yoon, D.J. Laser ultrasonic propagation imaging method in the frequency domain based on wavelet transformation. *Opt. Laser. Eng.* **2011**, *49*, 167–175. [CrossRef]

24. Michaels, T.E.; Michaels, J.E.; Ruzzene, M. Frequency-wavenumber domain analysis of guided wavefields. *Ultrason.* **2011**, *51*, 452–466. [CrossRef] [PubMed]

25. Rogge, M.D.; Leckey, C.A. Characterization of impact damage in composite laminates using guided wavefield imaging and local wavenumber domain analysis. *Ultrason* **2013**, *53*, 1217–1226. [CrossRef] [PubMed]

26. Park, B.; An, Y.-K.; Sohn, H. Visualization of hidden delamination and debonding in composites through noncontact laser ultrasonic scanning. *Compos. Sci. Technol.* **2014**, *100*, 10–18. [CrossRef]

27. An, Y.-K. Impact-induced delamination detection of composites based on laser ultrasonic zero-lag cross-correlation imaging. *Adv. Mater. Sci. Eng.* **2016**, *2016*, 6474852. [CrossRef]

28. Toyama, N.; Ye, J.; Kokuyama, W.; Yashiro, S. Non-contact ultrasonic inspection of impact damage in composite laminates by visualization of Lamb wave propagation. *Appl. Sci.* **2019**, *9*, 46. [CrossRef]

29. Ye, J.; Ito, S.; Toyama, N. Computerized ultrasonic imaging inspection: From shallow to deep learning. *Sensors* **2018**, *18*, 3820. [CrossRef]

30. Bendsøe, M.P.; Kikuchi, N. Generating optimal topologies in structural design using a homogenization method. *Comput. Methods Appl. Mech. Eng.* **1988**, *71*, 197–224. [CrossRef]
31. Bendsøe, M.P. Optimal shape design as a material distribution problem. *Struct. Optim.* **1989**, *1*, 193–202. [CrossRef]
32. Sigmund, O. A 99 line topology optimization code written in Matlab. *Struct. Multidiscip. Optim.* **2001**, *21*, 120–127. [CrossRef]
33. Zargham, S.; Ward, T.A.; Ramli, R.; Badruddin, I.A. Topology optimization: a review for structural designs under vibration problems. *Struct. Multidiscip. Optim.* **2016**, *53*, 1157–1177. [CrossRef]
34. Lee, J.S.; Kim, J.E.; Kim, Y.Y. Damage detection by the topology design formulation using modal parameters. *Int. J. Numer. Meth. Eng.* **2007**, *69*, 1480–1498. [CrossRef]
35. Nishizu, T.; Takezawa, A.; Kitamura, M. Eigenfrequecy-based damage identification method for non-destructive testing based on topology optimization. *Eng. Optim.* **2017**, *49*, 417–433. [CrossRef]
36. Reumers, P.; Van hoorickx, C.; Schevenels, M.; Lombaert, G. Density filtering regularization of finite element model updating problems. *Mech. Syst. Signal Pr.* **2019**, *128*, 282–294. [CrossRef]
37. Eslami, S.M.; Abdollahi, F.; Shahmiri, J.; Tavakkoli, S.M. Structural damage detection by using topology optimization for plane stress problems. *Int. J. Optim. Civil Eng.* **2019**, *9*, 159–176.
38. Niemann, H.; Morlier, J.; Shahdin, A.; Gourinat, Y. Damage localization using experimental modal parameters and topology optimization. *Mech. Syst. Signal Pr.* **2010**, *24*, 636–652. [CrossRef]
39. Shahdin, A.; Morlier, J.; Niemann, H.; Gourinat, Y. Correlating low energy impact damage with changes in modal parameters: diagnosis tools and FE validation. *Struct. Health Monit.* **2011**, *10*, 199–217. [CrossRef]
40. Morlier, J.; Niemann, H.; Shahdin, A. Topology optimization for robust damage localization using aggregated FRFs statistical criteria. *Appl. Mech. Mater.* **2014**, *629*, 513–518. [CrossRef]
41. Smith, W.D. A nonreflecting plane boundary for wave propagation problems. *J. Comput. Phys.* **1974**, *15*, 492–503. [CrossRef]
42. Bendsøe, M.P.; Sigmund, O. Material interpolation schemes in topology optimization. *Arch. Appl. Mech.* **1999**, *69*, 635–654. [CrossRef]
43. Hashin, Z.; Shtrikman, S. A variational approach to the theory of the elastic behaviour of multiphase materials. *J. Mech. Phys. Solids.* **1963**, *11*, 127–140. [CrossRef]

Article

Out-Of-Plane Permeability Evaluation of Carbon Fiber Preforms by Ultrasonic Wave Propagation

Francesca Lionetto *, **Francesco Montagna and Alfonso Maffezzoli**

Department of Engineering for Innovation, University of Salento, Via Arnesano, 73100 Lecce, Italy;
francesco.montagna@unisalento.it (F.M.); alfonso.maffezzoli@unisalento.it (A.M.)
* Correspondence: francesca.lionetto@unisalento.it; Tel.: +39-08-3229-7326

Received: 20 May 2020; Accepted: 10 June 2020; Published: 12 June 2020

Abstract: Out-of-plane permeability of reinforcement preforms is of crucial importance in the infusion of large and thick composite panels, but so far, there are no standard experimental methods for its determination. In this work, an experimental set-up for the measurement of unsaturated through thickness permeability based on the ultrasonic wave propagation in pulse echo mode is presented. A single ultrasonic transducer, working both as emitter and receiver of ultrasonic waves, was used to monitor the through thickness flow front during a vacuum assisted resin infusion experiment. The set-up was tested on three thick carbon fiber preforms, obtained by stacking thermal bonding of balanced or unidirectional plies either by automated fiber placement either by hand lay-up of unidirectional plies. The ultrasonic data were used to calculate unsaturated out-of-plane permeability using Darcy's law. The permeability results were compared with saturated out-of-plane permeability, determined by a traditional gravimetric method, and validated by some analytical models. The results demonstrated the feasibility and potential of the proposed set-up for permeability measurements thanks to its noninvasive character and the one-side access.

Keywords: ultrasound; time of flight; reinforcement; resin transfer molding (RTM); permeability; liquid composite molding; material characterization; composite manufacturing

---

1. Introduction

By combining the attractive features of different constituents, composite materials can be properly designed for specific applications, thus replacing traditional materials in an ever-increasing variety of products and applications in aerospace, automotive, marine, nanocomposite, construction, etc. [1–3]. More recently, the emerging trend towards lightweight, high performance, high functionality and high sustainability components is driving the research towards nanocomposites, multi-material hybrid structures and composites derived from renewable resources or from reusing approaches [4–9]. The growing demand of polymer matrix composites in several fields has also driven the research and development of new manufacturing processes. The challenge of manufacturing high quality and complex geometry components at relatively low cost and fast cycle times has pushed toward the development of liquid composite molding (LCM) processes [10,11]. These technologies usually involve the use of dry reinforcement preforms, an assembly of dry fiber layers that have been pre-shaped to the form of the desired product and bonded together using a binder resin, which are impregnated by resin injection or infusion into a mold [12,13]. Autoclave curing is usually used to limit the final void content, which remains a significant issue, related to the filling process, fiber wetting and tows impregnation. To achieve the required mechanical performance of the composite parts, multi-layered preforms with different architecture are generally used, such as unidirectional non crimped fabric and plain or twill woven fabric [14]. Furthermore, LCM requires low viscosity resins, which upon crosslinking would be brittle, being made of low molecular weight oligomers. Therefore, the higher molecular weight oligomers of the resin formulation, solid at room temperature, are added to the reinforcement as fiber coating or powders that are soluble in the matrix. These oligomers play also a role in preform fabrication, acting as low melting tackifying agent able to bind a ply to another upon heating [15]. The required performances are achieved in the final component only if a fast and complete filling of the preform with liquid resin is obtained thanks to the optimization of LCM process [16,17]. A suitable process design is necessary, which requires knowledge about different parameters affecting the filling behavior, such as mold geometry, resin viscosity, number and position of inlet and vent ports, etc. [18]. Besides the previous parameters, one of the most critical is the permeability of the reinforcement, defined as the resistance exhibited by the fibrous reinforcement against the resin flow [19]. This property determines the impregnation of the textile reinforcement with liquid resin and strongly depends on the reinforcement architecture, fiber orientation and volume fraction and on the procedure adopted to assemble and stabilize the plies into the preform [20]. The variability of textiles and the preforming operations affect the permeability of the fibrous reinforcements, resulting in possible unexpected or unwanted resin flow patterns [21]. Generally, permeability data are not available from the fabric manufacturer and must be determined depending on the fiber volume fraction. An accurate characterization of the reinforcement permeability is therefore fundamental to estimate the optimum process parameters for manufacturing high-quality components.

The permeability, K, of porous media is defined by Darcy's law [22] correlating the flow velocity V, pressure drop, ΔP, and viscosity, η, in a unidirectional flow over the length of the porous medium L:

$$V = -\frac{K}{\eta}\frac{\Delta P}{L} \tag{1}$$

The permeability of unidirectional preforms is an anisotropic property and is described by a second-order tensor [20]. Generally, by accounting for symmetry and assuming no coupling between in- and out-of-plane flow, only three components of the permeability tensor are not zero: (i) longitudinal in-plane permeability K_1, where in-plane refers to the reinforcement ply; (ii) transversal in-plane permeability K_2, oriented perpendicular to K_1 and lower than K_1; (iii) out-of-plane permeability K_3, oriented perpendicular to K_1 and K_2, lower than K_1 and of the same order of K_2 [21].

Different measurement methods of permeability are available in the literature [20,23,24]; each is capable to measure saturated or unsaturated permeability. Saturated permeability is a steady state permeability measured under a constant flow rate when the fiber reinforcement is fully saturated by the test fluid. Unsaturated permeability, also named unsteady-state or transient permeability, is measured when the reinforcement in the mold is progressively impregnated by a test fluid injected under a constant pressure [25]. However, there is a complete lack of standardization in the methods and experimental set-up, and the data, obtained using different methods and different fabric architecture, are not consistent [25]. Some benchmark studies demonstrated that when the same measuring procedure is used, the results are very different if different parameters are used, such as in terms of injection pressure and test fluid [26]. Most studies have been focused only on the in-plane permeability [27–29], disregarding the out-of-plane permeability, which is more difficult to measure accurately [30]. However, the out-of-plane (or through thickness) permeability is the dominant property in the infusion of large and flat panels with a high thickness when a resin distribution net is used. Therefore, its determination is of fundamental importance for the efficiency and robustness of LCM processes. The traditional video recording methods used for monitor flow front in RTM processes and determining in-plane permeability, cannot work well in the case of through-thickness permeability [31].

Several nondestructive methods have been investigated based on thermocouples [32], on the reflection at the flow front of electrical signals from thin metallic wires inside the fibers [33] guided mechanical waves [34], piezoelectric sensors [35] dielectric sensors [36] embedded in the fibers and pressure sensors positioned in the mold [37,38]. All of these measurement methods require both direct contact with the resin in order to detect the flow-front position or complete mold-filling and sensor embedded in the preform or into the mold [39,40]. The sensor embedding is not always allowed and is hence limited in application. Mounting sensors into the mold can affect the vacuum tightness of the mold and the sensors usually leave marks on the part surface so they can be only located at the border but not in the areas of interest [39].

Ultrasonic wave propagation has been widely recognized as a nondestructive testing (NDT) method applied to for the estimation of the physical and mechanical properties or the damage composite materials [41–45] and for process monitoring of composite materials [46,47]. Even the application of high intensity ultrasound in composite processing and joining present the potential for online monitoring [48–50]. Some research works proved the feasibility of using ultrasonic imaging systems for flow front monitoring even if the technique was limited to a single ply or thin preforms due to the severe attenuation of the ultrasound waves [31,51]. Stoven et al. [52] developed a set-up based on two ultrasonic probes operating in through transmission mode for flow monitoring and permeability determination in thickness direction. The ultrasonic based method presents several advantages: i) absence of any direct contact between the transducer and the composite and of any embedded sensor as the sound waves can be send through the mold wall; ii) low cost; iii) no disturbance to the fiber stack and liquid flow; iv) measurement of unsaturated permeability; v) reduced effort for preparing the experimental set-up; vi) no need of transparent tools; vii) compatible with all the fiber typologies [10,39,53].

Despite the numerous advantages, these ultrasonic based methods are still at a laboratory level due to the experience required to manage weak acoustic signals and the limitation of the resolution depth in thick preforms [53,54]. Moreover, the ultrasonic set-up described in the literature works in through transmission method with two ultrasonic transducers acting one as emitter and the other as receiver of ultrasonic waves [52], but sometimes the two-side access to the composite part is very difficult to achieve.

The aim of this work is to present a new experimental set-up for the measurement of unsaturated through thickness permeability based on the ultrasonic wave propagation in pulse echo mode, i.e., by using only one ultrasonic transducer working both as emitter and receiver of ultrasonic waves. Compared to the ultrasonic based systems previously reported in the literature, the proposed set-up enables one-side access, which is of great importance in composite manufacturing. This work stands out of the few previous related papers [10,52,53] by using the pulse echo technique with a single transducer, working both as emitter and receiver of ultrasonic waves, positioned at one side of the preform. The literature studies, in fact, use the ultrasonic transmission method, which is not always feasible when the composite is not accessible from both sides. The single ultrasonic probe, coupled to a vacuum assisted resin infusion (VARI) system, allows in-situ monitoring of unsteady preform impregnation. The system has been applied to multi-layered carbon fiber preforms. The robustness of the system has been tested on three different carbon fiber preforms used in aerospace field, each obtained by stacking balanced or unidirectional plies and then tested in vacuum infusion experiments. Saturated out-of-plane permeability has also been measured by a traditional gravimetric method and the results obtained by the two techniques have been compared. Finally, another novelty element is the validation of the permeability data, calculated from ultrasonic measurements, by some analytical models.

2. Materials and Methods

2.1. Materials

The preforms for permeability measurements were obtained using balanced and unidirectional carbon fiber fabrics and unidirectional (UD) tapes. The preforms investigated in this work are shown in Figure 1 and their properties are reported in Table 1.

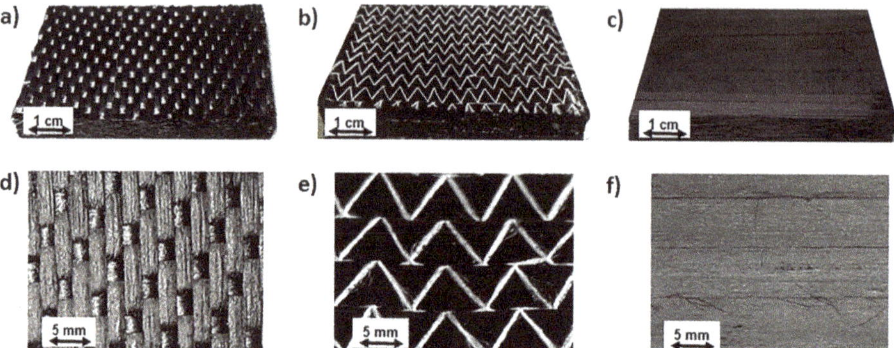

Figure 1. Preforms for permeability measurements: (**a,d**) balanced, (**b,e**) unidirectional (UD) stitched and (**c,f**) unidirectional (UD) consolidated by AFP.

Table 1. Carbon fiber preform configuration.

	A **Balanced Preform**	B **Stitched Preform**	C **AFP Preform**
Carbon fiber type	G0926 HS06K (Hexcel)	BNCF-24KIMS-(0)-196-600 (Cytec)	TX 1100 IMS65 24k(Cytec)
Fiber diameter (μm)	6.9	5.0	5.0
Fiber elastic modulus (GPa)	231	290	290
Fiber areal weight (g/m^2)	375	196	200
Number of layers	22	40	40
Nominal preform size (mm^3)	$80 \times 50 \times 8$	$80 \times 50 \times 8$	$80 \times 50 \times 8$
Preform manufacturing process	vacuum bagging	vacuum bagging	automated fiber placement

Preform A was a balanced fabric, produced by Hexcel (Stanford, CA, USA) with the trade name G0926 HS06K. It was a 0.38-mm-thick 5H satin with HEXTOW AS4C GP 6K with a nominal weight of 375 g/m^2, the same weight distribution in warp and weft directions and filled with an epoxy binder. The binder content was 4% of the weight of the carbon fiber [15]. Preforms A were obtained in laboratory by vacuum bagging of $[0]_{22}$ stacks of G0926 5H satin fabric in an oven during the following thermal cycle: heating up to 100 °C at 3 °C/min, isotherm for 75 min at 110 °C and cooling to room temperature at 1.5 °C/min. This cycle was able to dissolve the epoxy powders used as binder and to give stiffness to the preforms.

Preform B was a stitched unidirectional carbon fabric, produced by Solvay S.A. (Bruxelles, Belgium) with the trade name BNFC-24k IMS-(0)-196-600, obtained with IM65 carbon fibers containing a binder for preform fabrication. Preforms A and B, were provided by Leonardo SpA (Foggia, Italy), according to the same consolidation cycle.

Preform C was made of an unidirectional TX 1100 IMS65 24 k fabric, produced by Solvay S.A. (Bruxelles, Belgium) with IM65 carbon fibers. Preforms C, provided by Novotech Aerospace Advanced Technology srl (Avetrana, Italy), were prepared by automated fiber placement (AFP) using a tape width of 8 mm.

Permeability experiments were performed at room temperature using polyethylene glycol 400 (PEG400) as test fluid provided by Sigma-Aldrich (Milano, Italy). This latter has been chosen since its viscosity value at room temperature is in the viscosity range of a thermosetting resin used in resin infusion processes at high temperature. The viscosity of the test fluid was characterized by rheological analysis carried out in an ARES parallel plate rheometer with a 50-mm plate diameter (Rheometric Scientific).

2.2. Experimental Set-Up for Permeability Measurements by Ultrasonic Wave Propagation

The experimental set-up for out-of-plane permeability measurements by ultrasonic wave propagation is sketched not in scale in Figure 2. The carbon fiber (CF) preform was placed between two glass plates, each of them containing a hole for the inlet and outlet of the impregnation fluid. The thickness of the glass plates was 8 mm and 20 mm for the lower and upper plate, respectively. The higher thickness was chosen in order to avoid overlapping echoes due to the reflections from the glass plate and the wet carbon fiber preform. In order to provide the vacuum tightness of the measurement system, a silicone seal was used. A resin distribution net at the top and bottom side of the sample allowed a through-thickness unidirectional flow under a constant injection pressure, obtained by a Vacuum Assisted Resin Infusion (VARI). Fiber volume fraction was changed by adjusting the distance between the glass plates. After turning on the vacuum pump, the inlet valve was opened leading the fluid to flow from the tank through the preform. Since the measurement was carried out in unsteady conditions, an unsaturated permeability was determined. During the ultrasonic measurement, the balance—framed by a red dotted line in Figure 2—was not used.

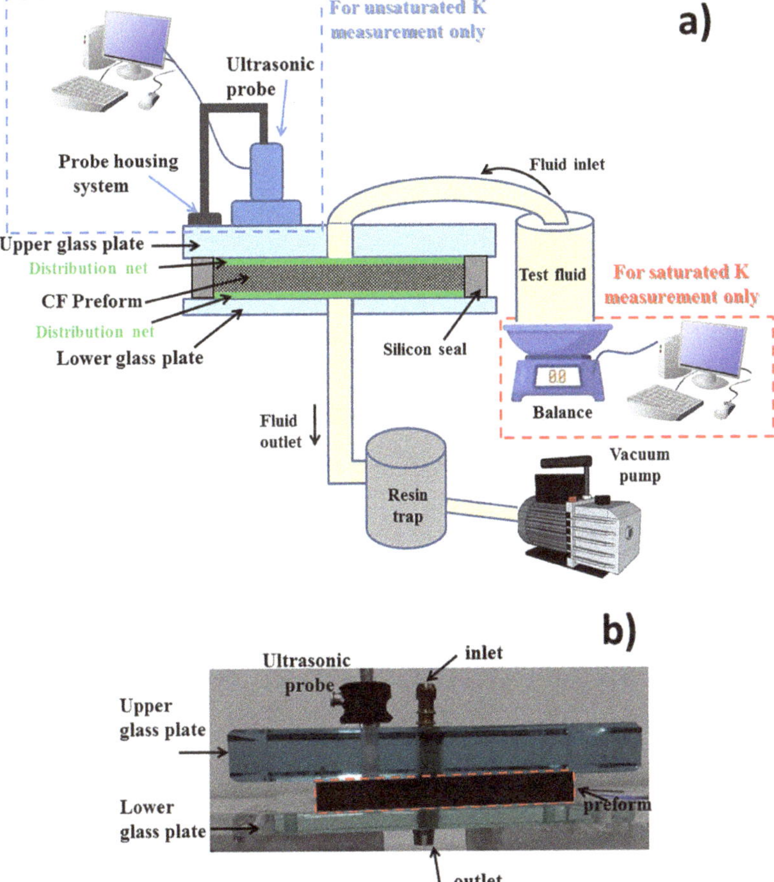

Figure 2. Experimental set-up for out-of-plane permeability measurements by ultrasonic wave propagation: (**a**) general sketch not in scale; (**b**) particular of the ultrasonic measurement device.

An ultrasonic transducer (A109-R Olympus Italia srl, Milano, Italy, center frequency =5 MHz, active diameter =13 mm) was fixed on the upper glass plate by a probe housing system capable to keep a constant force on the transducer. The ultrasonic frequency has been chosen as the best compromise among the wave attenuation and resolution, which increase with frequency, and the echo damping, which decreases with frequency and can lead to the overlapping of echoes related to different interfaces. An ultrasonic couplant gel was used to provide the transmission of sound energy between the transducer and the glass plate. The ultrasonic transducer, connected with a pulser-receiver (Sofratest, Ecquevilly, France), worked in pulse-echo mode, acting as emitter and receiver of ultrasonic waves at the same time [55]. The transducer was excited by a broadband pulser with an electric voltage of 200 V and a pulse duration of 200 nanoseconds. The echoes, sampled at a frequency of 60 MHz, were composed of 2048 points. The received signals, after digitalization, were automatically recorded each 0.15 sand displayed by a custom made software. Ultrasonic waves were continuously sent through the preform and it was possible to measure their time-of-flight that is the time taken for the wave to pass the preform and come back to the transducer after reflection at interfaces characterized by different acoustic impedances [56]. The two first round-trip echoes in the wet preform were considered. The ultrasonic measurement device has been set-up for carbon fiber preforms up to 10 mm. However, thicker preforms can be investigated by increasing the thickness of the upper glass plate.

2.3. Experimental Set-Up for Measurement of Saturatedpermeability

The same experimental set-up sketched in Figure 2 was used also for the measurement of saturated permeability. In this case, the ultrasonic probe was not used since a gravimetric measurement method was adopted. A becher filled with the test fluid was placed on a balance and, after turning on the vacuum pump, the inlet valve was opened leading the fluid to flow through the preform. Out-of-plane saturated permeability was measured after achieving steady state conditions. When the preform was saturated with the fluid, the quantity of fluid that entered from the inlet was equal to the quantity that came out from the outlet. A constant flow rate, measured by gravimetric method, under a constant pressure generated by vacuum was produced across the thickness of the different preforms. During the test, the weight of the fluid was measured as a function of time using a balance with a precision of 10^{-4} g.

3. Results

3.1. Rheological Analysis of the Model Fluid

As shown in Figure 3a, PEG400 presents a Newtonian behavior, indicating that its viscosity is not dependent on the shear rate. The temperature dependence of the viscosity has been modeled by an Arrhenius like equation, a common approach for oligomers and polymers at a temperature much higher than their T_g [57]:

$$\eta = \eta_0 \exp\left(\frac{E_a}{RT}\right) \tag{2}$$

where η_0 is a pre-exponential factor, T the absolute temperature, R the universal gas constant and E_a the activation energy. The nonlinear fitting of PEG 400 viscosity with Equation (2) (see Figure 3b) provides an E_a value of 36.9 kJ/mol. Depending on the temperature of the infusion, Equation (2) has been used to determine the value of viscosity needed for permeability calculation from Darcy's law.

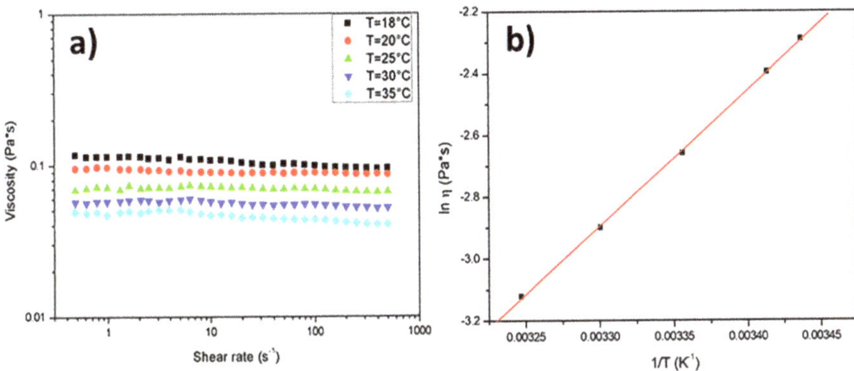

Figure 3. (**a**) Effect of the shear rate on the viscosity of the test fluid at different temperatures; (**b**) temperature dependence of the viscosity of the test fluid.

3.2. Saturated Out-Of-Plane Permeability Measurements by VARI Process

Saturated out-of-plane permeability has been determined from gravimetric measurements at constant flow rate during a Vacuum Assisted Resin Infusion (VARI) process [20]. When steady state conditions are achieved, the quantity of the fluid that enters the preform is equal to the quantity that comes out per time unit. The decrease in weight of the fluid is then monitored as a function of the time. The out-of-plane saturated permeability $K_{3\text{-sat}}$ is thus obtained by plotting the fluid weight as a function of time, t, according to the following equation:

$$QAt = \text{Weight} = \frac{\rho K_{3-sat} A}{\eta} \frac{\Delta P}{L} t \tag{3}$$

where Q is the flow rate, A is the flow channel cross-sectional area, ρ is the fluid density, ΔP is the pressure difference (the pressure drop in the pipes was neglected, as well as the effect of gravity), L is the specimen thickness, respectively. The fluid viscosity η has been determined according to Equation (2), accounting for the test temperature. A typical plot of the weight loss as a function of time is reported in Figure 4a for Preform A.

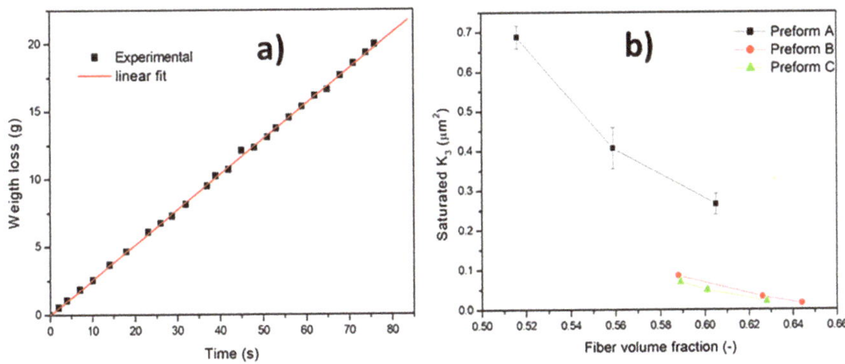

Figure 4. (**a**) Weight loss as a function of time for preform A; (**b**) saturated out-of-plane permeability as a function of fiber volume fraction.

Considering that the preform is placed between two glass plates at a cavity height h, the fiber volume fraction of the preform can be obtained according to following equation:

$$V_f = \frac{nS_0}{\rho_f h} \tag{4}$$

where n is the number of plies, ρ_f is the fiber density and S_0 the areal weight of a single ply.

The effect of fiber volume fraction on the saturated out-of-plane permeability of all the investigated preforms is reported in Figure 4b. The values are the average of at least three measurements. As expected, by increasing the fiber volume fraction, K_3 decreases. The same reinforcement can be characterized by a different permeability depending on the fiber content, related to the cavity height. The different K_3 values among the different preform typologies can be attributed to the different structure of the materials. Preform A is a woven UD fabric with stabilizing weft tows, while the Preform B and C are unidirectional, the former stabilized by stitches, the latter, used in AFP, is a true UD tape. Comparing preforms A and B, both obtained by vacuum bagging, the same pressure produced a higher compaction of fibers for preform B than for preform A, thus leading to a higher fiber volume fraction. The dependence of permeability on fiber volume fraction is different for preform A since it is more compressible than the other two as a consequence of the presence of a woven fabric in it.

The measured saturated out-of-plane permeabilities are also reported in Table 2. To the best of author knowledge, a comparison with literature data can be only made between the values of unidirectional Preform C and the work of Aziz et al. [58] who measured $K_{3\text{-sat}}$ by a saturated unidirectional flow device on TX 1100 quasi-isotropic preforms produced by automated fiber placement with different band widths. In particular, at a fiber content of 58%, the obtained saturated permeability was 0.0831 μm^2 and 0.0185 μm^2 on preforms with a band width of 6.35 mm and 12.7 mm, respectively. The experimental value of 0.702 μm^2 obtained in this study on Preform C (produced by automated fiber placement with a band width of 8 mm) at $V_f = 58.8\%$ is in the range measured by Aziz et al. [58], at the same preform material and fiber content, even if the preform architecture is different.

Table 2. Out-of-plane saturated and unsaturated permeabilities for the investigated preforms.

Material	V_f (%)	Out-Of-Plane Saturated Permeability (μm^2)	Out-Of-Plane Unsaturated Permeability (μm^2)
	51.6	0.688 ± 0.0290	0.306 ± 0.032
Preform A	55.9	0.407 ± 0.0052	0.143 ± 0.039
	60.5	0.266 ± 0.0027	0.107 ±0.029
	58.8	0.087 ± 0.005	0.057 ± 0.009
Preform B	62.6	0.034 ± 0.006	0.035 ± 0.009
	64.4	0.017 ± 0.005	0.018 ± 0.009
	58.9	0.070 ± 0.010	0.051 ± 0.003
Preform C	60.1	0.050 ± 0.007	0.040 ± 0.006
	62.8	0.021 ± 0.005	0.029 ± 0.004

3.3. Out-Of-Plane Permeability Measurements by Ultrasonic Wave Propagation

Unsaturated K_3 permeability has been measured using a single ultrasonic transducer working in pulse-echo mode, i.e., working either as emitter either as receiver. A dedicated software is able to visualize and save the echoes and to record the time delay between them (time of flight). The principle underlying the measurement is based on the reflection of the ultrasonic wave at the interface between two materials of different density and elastic properties [59]. Since ultrasounds are almost completely reflected at the interface between a solid or a liquid and air, and strongly attenuated by scattering in porous media, the reflected ultrasonic signal (echo) from a dry preform is nearly negligible while that one from a wetted preform is clearly detectable.

As sketched in Figure 5a, when the infusion has not yet begun and the carbon fiber (CF) preform is still dry, the ultrasonic wave is almost completely reflected at the glass/preform interface because ultrasound is very scarcely transmitted in air. Therefore, in Figure 5a, only one ultrasonic wave path is sketched by green arrows. The corresponding echogram (Figure 5b) reports the signals relative to the multiple reflections back and forth from the upper glass plate. No echo relative to reflections from the preform is observed.

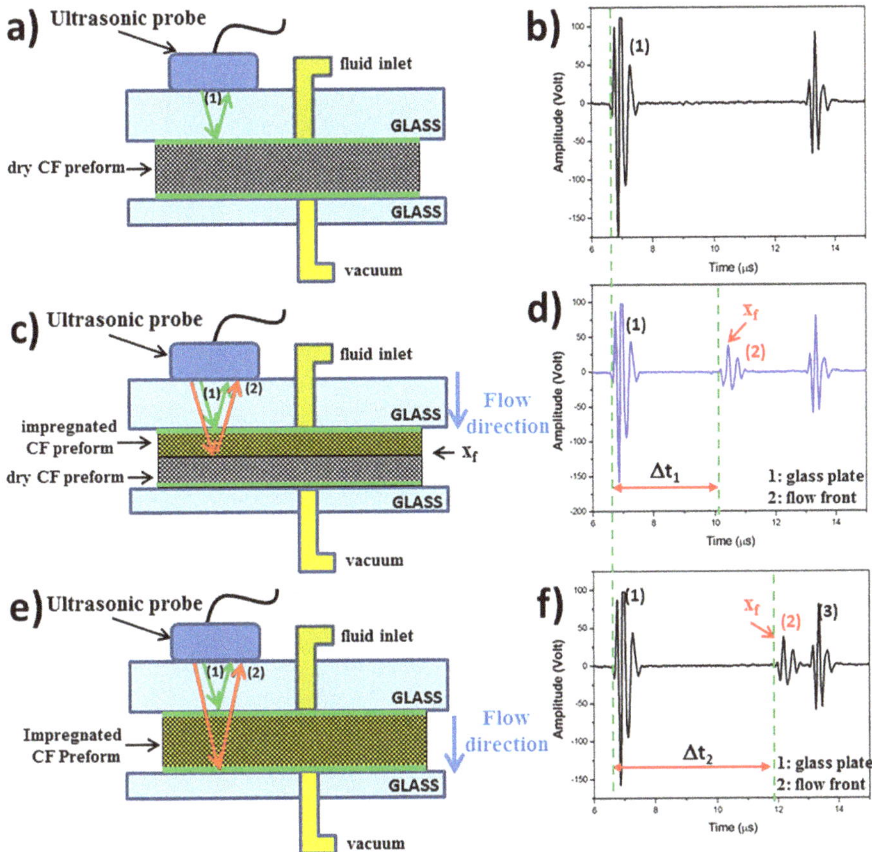

Figure 5. Ultrasonic wave path and corresponding echograms at different stages of the impregnation process: (**a**,**b**) no impregnation; (**c**,**d**) partial impregnation;(**e**,**f**) complete impregnation.

As sketched in Figure 5c, after the beginning of infusion, when the preform is partially impregnated, a fraction of the ultrasonic wave is transmitted at the glass/wet preform interface and travels through the wet preform but it is totally reflected (red arrows) at the flow front, i.e., at the interface between the wet preform and the dry preform. The signals relative to the reflections at the glass/wet preform interface (echo No.1) and at the wet/dry preform interface (echo No.2) are shown in Figure 5d. The time difference, Δt, between the two peak times relative to the echo from the partial impregnated preform (No.2) and the echo from the glass plate (No.1) is called time of flight (TOF)and corresponds to the time the ultrasonic waves take to travel twice between the glass plate and the impregnation front.

When the preform is filled, the path of ultrasonic wave is represented by the red arrows in Figure 5e. Consequently, the echo No.2 in Figure 5f is right shifted while the position of the two echoes

No.1 and No.3 relative to multiple reflections at the top glass plate-preform interface remains constant. Therefore, the recorded time of flight of the wave reflected from preform increases since the distance of the flow front from the transducer increases. As observable from Figure 5d–f, the TOF of partial impregnated preform Δt_1 is lower than that of completely impregnated preform Δt_2. The echo No.2 can be therefore used to monitor the flow front position through the thickness of the preform.

A custom-made software has been used to record the TOF during the infusion of the carbon fiber preform. The time of flight measured during an infusion experiment is reported as a function of the infusion time in Figure 6a. TOF increases when the infusion starts, then it reaches a plateau value when the preform is fully impregnated. At this point, it can be deduced that the impregnation of the preform is complete.

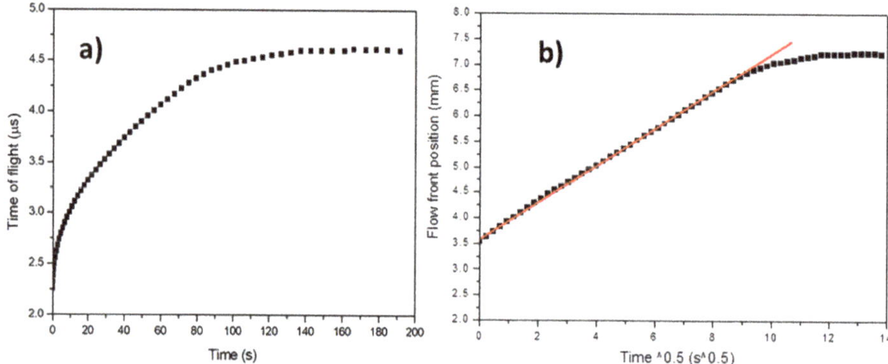

Figure 6. Preform C at fiber volume fraction of 0.628: (**a**) measured ultrasonic time of flight during the infusion; (**b**) calculated flow front position from ultrasonic data.

The TOF value can be used to determine the distance x_f of the flow front from the glass plate, which in the pulse echo mode is given by:

$$x_f = \frac{\Delta t c}{2} \tag{5}$$

where Δt is the TOF value and c the longitudinal wave velocity in the wet preform. This latter has been determined for each analyzed preform at each fiber volume content, by applying the inverse rule of mixture for a composite material with carbon fibers and PEG400 matrix:

$$c = \frac{1}{\frac{V_f}{c_f} + \frac{V_m}{c_m}} \tag{6}$$

where V_f and V_m are the volume fraction of carbon fiber and PEG matrix, respectively, while c_f and c_m are the sound velocity of carbon fiber and PEG matrix, respectively. The sound velocity of PEG 400 has been measured and it is equal to 1507 m/s. The sound velocity of carbon fiber has been estimated from the elastic modulus and density of the different fibers used for the three preforms according to the following equation:

$$c_f = \sqrt{\frac{E}{\rho}} \tag{7}$$

where ρ is the density, equal to 1.78 g/cm^3 for all the fibers, while the elastic modulus E is equal to 231 GPa for the carbon fibers of preform A and 290 GPa for those of preform B and C (see Table 1). A sound speed of 11,392 m/s for carbon fibers of preform A and 12,764 m/s for carbon fibers of preform B and C has been obtained from Equation (7). The ultrasonic velocity value c, calculated according to Equation (6), is considered constant during the experiment. The eventual presence of microscopic air bubbles during the flow does not affect ultrasonic velocity but can decreases the amplitude of the ultrasonic wave [46].

The x_f values as a function of square root of time, reported in Figure 6b, present an initial linear behavior which can be used for the determination of unsaturated out-of-plane permeability $K_{3-unsat}$ of the preform using the Darcy's equation, valid in the case of one-dimensional flow and constant injection pressure:

$$x_f = \sqrt{\frac{2K_{3-unsat}\Delta P}{\eta}}\,\sqrt{t_f} \tag{8}$$

where ΔP is the pressure difference and η the viscosity of the fluid, determined according to Equation (2), accounting for the test temperature.

The slope obtained from the linear fit of the data can be used for the determination of $K_{3-unsat}$, according to the following equation:

$$K_{3-unsat} = \frac{(slope)^2\eta}{2\Delta P} \tag{9}$$

The values of saturated and unsaturated permeabilities of all the investigated preforms are reported in Table 2. The values are the average of at least three measurements. For each preform, as the fiber volume fraction increases, the permeability values decrease, due to the increased fiber compaction that leads to a reduced flow rate under the same pressure gradient. Moreover, at the same fiber volume fraction and for the same preform, the saturated out-of-plane permeability is higher than the unsaturated out-of-plane permeability. The discrepancy between saturated and unsaturated permeabilities, which has been observed also for in-plane permeability, has been explained by many authors by the void formation during the unsaturated flow. During the filling process, the liquid flow advancement is not uniform and air is entrapped at the flow front, creating partially impregnated zone leading to a change of the hydraulic conductivity [25].

The effect of fiber volume fraction on the unsaturated out-of-plane permeability of all the investigated preforms is reported in Figure 7. Preform A is characterized by higher permeability at similar V_f, between 0.585 and 0.605, probably as a consequence of the different architecture of each ply. The weft tows in preform A, not only limit its compressibility leading to a lower V_f at the same compaction level but are also responsible of the presence of lower resistance pathways to fluid advancement. On the other hand, lower permeability of preforms B and C at the same V_f, can be attributed to the absence of low resistance pathways. The slightly higher permeability of preform B compared to C is probably due to the stitching yarns used in preform B to stabilize the UD arrangement (Figure 1). The impregnation experiments made for the measurement of in-plane permeabilities K_1 and K_2 (not reported), indicated that these yarns are more easily wetted by the fluid which find a low resistance pathways inside them [60]. On the other hand, the carbon fiber tows of preform C, obtained using tapes stabilized by a binder also acting as tackifying agent during the lay up, are better compacted, even if the control of temperature and pressure applied during AFP is less accurate than in vacuum bagging.

A comparison with literature data is very difficult since the reported permeability data are characterized by a strong scattering depending on the measurement method, the fiber preform, the test fluid and the fluid injection parameters. As also observed also by Konstantopoulos et al. [53], the literature data can be used only as a reference to confirm that the results of this study lie in the correct order of magnitude: the unsaturated data obtained by Agougue et al. [30], on TX100 preform at V_f=58% are very close to the results obtained in this work. Despite the different preform architecture (quasi-isotropic for [30] and unidirectional in this study) and the different measurement method, the same order of magnitude of 100 μm^2 is obtained by correcting the value reported by Agogue et al. [30] by the (1-V_f) coefficient.

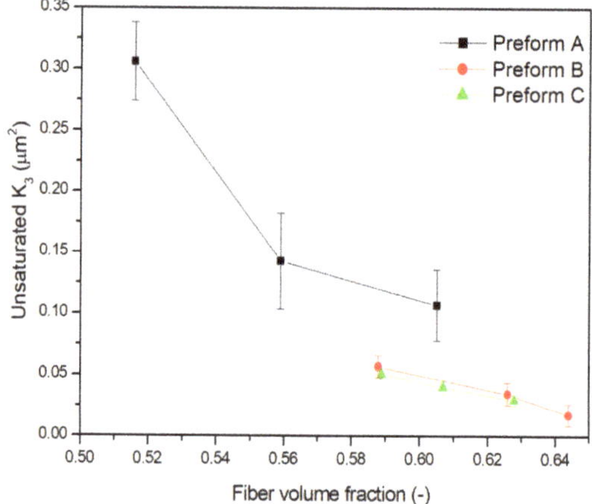

Figure 7. Unsaturated out-of-plane permeability as a function of fiber volume fraction.

3.4. Mathematical Modeling

The values of out-of-plane permeability, both saturated and unsaturated, of the three different preforms have been further validated by comparison with the results of predictive models. As recently reported by Karaki et al. [18,61], among the several mathematical models available in the literature for estimating the permeability of unidirectional reinforcements, the most accurate ones, able to fit the experimental data of out-of-plane permeability of unidirectional yarns, are those proposed by Gebart [62] and Berdichevsky and Cai [63].

Gebart [62] derived an analytical model to predict the unidirectional permeability starting from Navier-Stokes equation, considering two fibers arrangements, a quadratic array and a hexagonal array:

$$K_3 = C_1 \left(\sqrt{\frac{V_{fmax}}{V_f}} - 1 \right)^{\frac{5}{2}} r_f^2 \tag{10}$$

where V_f is the fiber volume fraction, r_f the fiber radius and the constants C_1 and V_{fmax} depend on the fiber packing arrangement according to the values reported in Table 3.

Table 3. Gebart model: constant values for quadratic and hexagonal fiber arrangement.

Fiber Arrangement	C_1	V_{fmax}
Quadratic	$\frac{16}{9\pi \times \sqrt{2}}$	$\frac{\pi}{4}$
Hexagonal	$\frac{16}{9\pi \times \sqrt{6}}$	$\frac{\pi}{2 \times \sqrt{3}}$

Berdichesvky and Cai [63] derived an unified empirical model where Stokes flow and Darcy flow are considered at different regions and the gap between neighboring fibers governs the flow resistance:

$$K_3 = 0.229 r^2 \left(\frac{1.814}{V_a} - 1 \right) \left(\frac{\left(1 - \sqrt{\frac{V_f}{V_a}} \right)}{\sqrt{\frac{V_f}{V_a}}} \right)^{2.5} \tag{11}$$

where V_f is the fiber volume fraction and V_a is equal to 0.9069 and 0.7854 in case of hexagonal or quadratic fiber arrangement, respectively.

Models predictions and the experimental results on the unsaturated and saturated permeabilities are compared in Figure 8, where only the best fit among modeling and experimental data are reported. A nonlinear fit of the K_3 values as a function of fiber volume fraction has been performed using Equations (10) and (11), in order to obtain the fiber radius r_f, whose values are reported in Table 4. The quadratic fiber arrangement is not able to properly model the experimental values, as inferred by the high differences reported in Table 4, since the obtained fiber radius is higher than the nominal value, which is equal to 3.35 µm for Preform A and 2.5 µm for Preforms B and C. Therefore, the fit obtained with the quadratic fiber arrangement with both the models are not reported in Figure 8 for the sake of clarity.

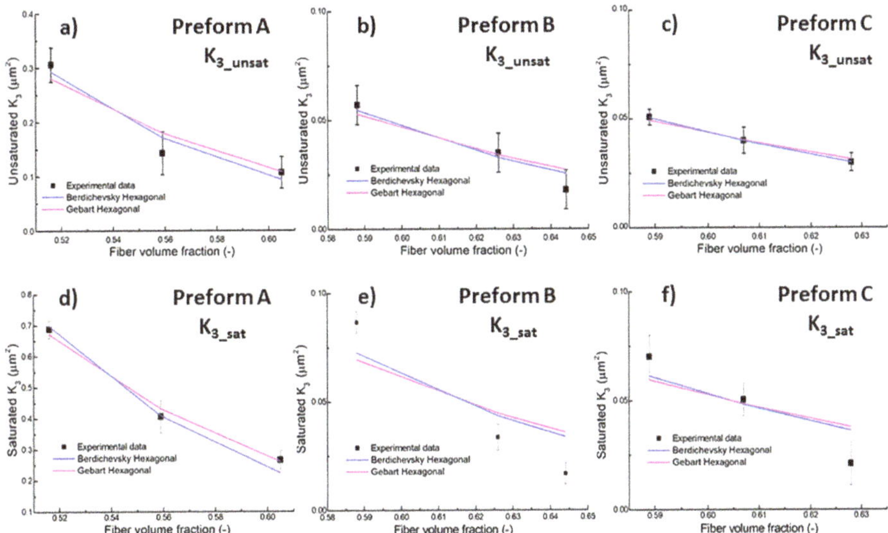

Figure 8. Mathematical modeling of unsaturated (**a–c**) and saturated permeability (**d–f**) values: (**a,d**) preform A; (**b,e**) preform B; (**c,f**) preform C.

On the other hand, the best fit is always obtained adopting the hexagonal array and Berdichevsky and Cai model, with differences from the nominal fiber radius lower than 15%. Only the best fit of saturated K_3 of preform A, with all models, leads always to fiber radii much larger than the nominal ones. From Figure 8 it can be observed good agreement among the unsaturated K_3 values calculated by ultrasonic measurement and the model prediction. This can be considered a further validation of the proposed measurement method, based on ultrasonic wave propagation.

It should be also underlined that the schematization in hexagonal and quadratic array is often ideal since the fiber distribution is often irregular and falls in between hexagonal and quadratic arrangement. Fiber clustering, in real preforms, makes reasonable the almost always higher radii obtained from fitting compared to the nominal ones.

Table 4. Unsaturated out-of-plane fiber radius obtained from the mathematical modeling of the experimental K_3 values and percentage difference from the nominal fiber radius.

Material	Out-Of-Plane Permeability	Model	Fiber Radius from Model Best Fit, r_f (μm)	Nominal Fiber Radius r_{fn} (μm)	Difference $[(r_{fn} - r_f)/r_f] \times 100$ (%)
Preform A	Unsaturated	Gebart-hexagonal	4.57	3.45	+32
		Gebart-quadratic	5.45	3.45	+58
		Berdichevsky-hexagonal	3.23	3.45	−6.8
		Berdichevsky-quadratic	4.80	3.45	+39
	Saturated	Gebart-hexagonal	7.09	3.45	+105
		Gebart-quadratic	8.41	3.45	+144
		Berdichevsky-hexagonal	4.99	3.45	+45
		Berdichevsky-quadratic	7.39	3.45	+114
Preform B	Unsaturated	Gebart-hexagonal	2.89	2.5	+16
		Gebart-quadratic	4.04	2.5	+62
		Berdichevsky-hexagonal	2.20	2.5	−12
		Berdichevsky-quadratic	3.73	2.5	+49
	Saturated	Gebart-hexagonal	3.32	2.5	+33
		Gebart-quadratic	4.69	2.5	+88
		Berdichevsky-hexagonal	2.53	2.5	+1.2
		Berdichevsky-quadratic	4.43	2.5	+77
Preform C	Unsaturated	Gebart-hexagonal	2.81	2.5	+12
		Gebart-quadratic	3.87	2.5	+55
		Berdichevsky-hexagonal	2.13	2.5	−15
		Berdichevsky-quadratic	3.65	2.5	+46
	Saturated	Gebart-hexagonal	3.09	2.5	+24
		Gebart-quadratic	4.29	2.5	+72
		Berdichevsky-hexagonal	2.35	2.5	−6
		Berdichevsky-quadratic	4.05	2.5	+62

4. Conclusions

An experimental set-up for the measurement of unsaturated through thickness permeability based on the ultrasonic wave propagation in pulse echo mode was proposed. A single ultrasonic transducer, working both as emitter and receiver of ultrasonic waves, was used to monitor the through thickness flow front during a vacuum assisted resin infusion set-up. The new measurement method was tested on three thick carbon fiber preforms, obtained by stacking and vacuum bagging or by automated fiber placement of unidirectional plies. The ultrasonic data were used to calculate unsaturated out-of-plane permeability applying Darcy's law. The permeability results were compared with saturated out-of-plane permeability, determined by a traditional gravimetric method.

For each preform, as the fiber volume fraction increased, the permeability values decreased, due to the increased fiber compaction that lead to a slowdown of the fluid flow. Moreover, at the same fiber volume fraction and for the same preform material, the saturated out-of-plane permeabilities were higher than the unsaturated out-of-plane permeabilities.

Finally, the permeability data were validated by two analytical models: Berdichevsky and Cai model and Gebart model. The best agreement was obtained using Berdichevsky and Cai model assuming a hexagonal fiber arrangement.

Author Contributions: Conceptualization, F.L. and A.M.; Methodology, F.L. and A.M.; Experimental activity, F.L. and F.M.; Data Curation, F.L. and F.M.; Writing, Review and Editing, F.L. and A.M.; Supervision, A.M.; Funding Acquisition, A.M. All authors have read and agreed to the published version of the manuscript.

Funding: The research leading to these results was partially funded from by the Italian Minister for University and Research (MIUR) through the PON research project SPIA.

Acknowledgments: The authors kindly acknowledge Silvio Pappadà from CETMA, Giuseppe Totaro from Leonardo S.p.A. and Giuseppe Barile from Novotech Aerospace Advanced Technology srl for providing the three different carbon fiber preforms. Riccardo Dell'Anna and Alessia Peluso are kindly acknowledged for the assistance in the ultrasonic set-up and permeability measurements.

Conflicts of Interest: The authors declare no conflict of interest. The funders had no role in the design of the study; in the collection, analyses, or interpretation of data; in the writing of the manuscript, or in the decision to publish the results.

References

1. Donadei, V.; Lionetto, F.; Wielandt, M.; Offringa, A.; Maffezzoli, A. Effects of Blank Quality on Press-Formed PEKK/Carbon Composite Parts. *Materials* **2018**, *11*, 1063. [CrossRef] [PubMed]
2. De Bruijn, T.; Vincent, G.A.; Meuzelaar, J.; Nunes, J.P.; van Hattum, F. Design, manufacturing and testing of a rotorcraft access panel door from recycled carbon fiber reinforced polyphenylenesulfide. In Proceedings of the SAMPE North America Conference and Exhibition, Charlotte, NC, USA, 20–23 May 2019.
3. Paleo, A.J.; Zille, A.; Van Hattum, F.W.; Ares-Pernas, A.; Moreira, J.A. Dielectric relaxation of near-percolated carbon nanofiber polypropylene composites. *Phys. B Condens. Matter* **2017**, *516*, 41–47. [CrossRef]
4. Barra, G.; Guadagno, L.; Vertuccio, L.; Simonet, B.; Santos, B.; Zarrelli, M.; Arena, M.; Viscardi, M. Different Methods of Dispersing Carbon Nanotubes in Epoxy Resin and Initial Evaluation of the Obtained Nanocomposite as a Matrix of Carbon Fiber Reinforced Laminate in Terms of Vibroacoustic Performance and Flammability. *Materials* **2019**, *12*, 2998. [CrossRef] [PubMed]
5. Arena, M.; Viscardi, M.; Barra, G.; Vertuccio, L.; Guadagno, L. Multifunctional performance of a nano-modified fiber reinforced composite aeronautical panel. *Materials* **2019**, *12*, 869. [CrossRef] [PubMed]
6. Martulli, L.M.; Creemers, T.; Schöberl, E.; Hale, N.; Kerschbaum, M.; Lomov, S.V.; Swolfs, Y. A thick-walled sheet moulding compound automotive component: Manufacturing and performance. *Compos. Part A Appl. Sci. Manuf.* **2020**, *128*, 105688. [CrossRef]
7. Schneeberger, C.; Wong, J.C.H.; Ermanni, P. Hybrid bicomponent fibres for thermoplastic composite preforms. *Compos. Part A Appl. Sci. Manuf.* **2017**, *103*, 69–73. [CrossRef]
8. Luzi, F.; Torre, L.; Kenny, J.M.; Puglia, D. Bio-and fossil-based polymeric blends and nanocomposites for packaging: Structure–property relationship. *Materials* **2019**, *12*, 471. [CrossRef]
9. Greco, A.; Lionetto, F.; Maffezzoli, A. Processing and characterization of amorphous polyethylene terephthalate fibers for the alignment of carbon nanofillers in thermosetting resins. *Polym. Compos.* **2015**, *36*, 1096–1103. [CrossRef]
10. Becker, D.; Grössing, H.; Konstantopoulos, S.; Fauster, E.; Mitschang, P.; Schledjewski, R. An evaluation of the reproducibility of ultrasonic sensor-based out-of-plane permeability measurements: A benchmarking study. *Adv. Manuf. Polym. Compos. Sci.* **2016**, *2*, 34–45. [CrossRef]
11. Semperger, O.V.; Suplicz, A. The Effect of the Parameters of T-RTM on the Properties of Polyamide 6 Prepared by in Situ Polymerization. *Materials* **2020**, *13*, 4. [CrossRef]
12. Parnas, R.S. *Liquid Composite Molding*; Carl Hanser Verlag GmbH Co KG: Munich, Germany, 2014; ISBN 3446443029.
13. Mallick, P.K. Thermoset-matrix composites for lightweight automotive structures. In *Materials, Design and Manufacturing for Lightweight Vehicles*; Elsevier: Amsterdam, The Netherlands, 2010; pp. 208–231.
14. Seong, D.G.; Kim, S.; Lee, D.; Yi, J.W.; Kim, S.W.; Kim, S.Y. Prediction of Defect Formation during Resin Impregnation Process through a Multi-Layered Fiber Preform in Resin Transfer Molding by a Proposed Analytical Model. *Materials* **2018**, *11*, 2055. [CrossRef] [PubMed]
15. Lionetto, F.; Moscatello, A.; Maffezzoli, A. Effect of binder powders added to carbon fiber reinforcements on the chemoreology of an epoxy resin for composites. *Compos. Part B Eng.* **2017**, *112*, 243–250. [CrossRef]
16. Sun, Z.; Xiao, J.; Tao, L.; Wei, Y.; Wang, S.; Zhang, H.; Zhu, S.; Yu, M. Preparation of High-Performance Carbon Fiber-Reinforced Epoxy Composites by Compression Resin Transfer Molding. *Materials* **2019**, *12*, 13. [CrossRef] [PubMed]
17. Geng, Z.; Yang, S.; Zhang, L.; Huang, Z.; Pan, Q.; Li, J.; Weng, J.; Bao, J.; You, Z.; He, Y. Self-Extinguishing Resin Transfer Molding Composites Using Non-Fire-Retardant Epoxy Resin. *Materials* **2018**, *11*, 2554. [CrossRef]
18. Karaki, M.; Younes, R.; Trochu, F.; Lafon, P. Progress in Experimental and Theoretical Evaluation Methods for Textile Permeability. *J. Compos. Sci.* **2019**, *3*, 73. [CrossRef]
19. Luo, Y.; Verpoest, I.; Hoes, K.; Vanheule, M.; Sol, H.; Cardon, A. Permeability measurement of textile reinforcements with several test fluids. *Compos. Part A Appl. Sci. Manuf.* **2001**, *32*, 1497–1504. [CrossRef]

20. Naik, N.K.; Sirisha, M.; Inani, A. Permeability characterization of polymer matrix composites by RTM/VARTM. *Prog. Aerosp. Sci.* **2014**, *65*, 22–40. [CrossRef]
21. Sharma, S.; Siginer, D.A. Permeability measurement methods in porous media of fiber reinforced composites. *Appl. Mech. Rev.* **2010**, *63*. [CrossRef]
22. Darcy, H. *Les Fontaines Publiques de la Ville de Dijon: Exposition et Application*; Victor Dalmont: Paris, France, 1856.
23. Parnas, R.S.; Howard, J.G.; Luce, T.L.; Advani, S.G. Permeability characterization. Part 1: A proposed standard reference fabric for permeability. *Polym. Compos.* **1995**, *16*, 429–445. [CrossRef]
24. Verheus, A.S.; Peeters, J.H.A. The role of reinforcement permeability in resin transfer moulding. *Compos. Manuf.* **1993**, *4*, 33–38. [CrossRef]
25. Park, C.H.; Krawczak, P. Unsaturated and Saturated Permeabilities of Fiber Reinforcement: Critics and Suggestions. *Front. Mater.* **2015**, *2*, 38. [CrossRef]
26. Vernet, N.; Ruiz, E.; Advani, S.; Alms, J.B.; Aubert, M.; Barburski, M.; Barari, B.; Beraud, J.M.; Berg, D.C.; Correia, N. Experimental determination of the permeability of engineering textiles: Benchmark II. *Compos. Part A Appl. Sci. Manuf.* **2014**, *61*, 172–184. [CrossRef]
27. Amico, S.; Lekakou, C. An experimental study of the permeability and capillary pressure in resin-transfer moulding. *Compos. Sci. Technol.* **2001**, *61*, 1945–1959. [CrossRef]
28. Ding, L.; Shih, C.; Liang, Z.; Zhang, C.; Wang, B. In situ measurement and monitoring of whole-field permeability profile of fiber preform for liquid composite molding processes. *Compos. Part A Appl. Sci. Manuf.* **2003**, *34*, 779–789. [CrossRef]
29. May, D.; Aktas, A.; Advani, S.G.; Berg, D.C.; Endruweit, A.; Fauster, E.; Lomov, S.V.; Long, A.; Mitschang, P.; Abaimov, S. In-plane permeability characterization of engineering textiles based on radial flow experiments: A benchmark exercise. *Compos. Part A Appl. Sci. Manuf.* **2019**, *121*, 100–114. [CrossRef]
30. Agogue, R.; Chebil, N.; Deleglise-Lagardere, M.; Beauchene, P.; Park, C.H. Efficient permeability measurement and numerical simulation of the resin flow in low permeability preform fabricated by automated dry fiber placement. *Appl. Compos. Mater.* **2018**, *25*, 1169–1182. [CrossRef]
31. Thomas, S.; Bongiovanni, C.; Nutt, S.R. In situ estimation of through-thickness resin flow using ultrasound. *Compos. Sci. Technol.* **2008**, *68*, 3093–3098. [CrossRef]
32. Tuncol, G.; Danisman, M.; Kaynar, A.; Sozer, E.M. Constraints on monitoring resin flow in the resin transfer molding (RTM) process by using thermocouple sensors. *Compos. Part A Appl. Sci. Manuf.* **2007**, *38*, 1363–1386. [CrossRef]
33. Buchmann, C.; Filsinger, J.; Ladstätter, E. Investigation of Electrical Time Domain Reflectometry for infusion and cure monitoring in combination with electrically conductive fibers and tooling materials. *Compos. Part B Eng.* **2016**, *94*, 389–398. [CrossRef]
34. Visvanathan, K.; Balasubramaniam, K. Ultrasonic torsional guided wave sensor for flow front monitoring inside molds. *Rev. Sci. Instrum.* **2007**, *78*, 15110. [CrossRef]
35. Tuloup, C.; Harizi, W.; Aboura, Z.; Meyer, Y.; Ade, B.; Khellil, K. Detection of the key steps during Liquid Resin Infusion manufacturing of a polymer-matrix composite using an in-situ piezoelectric sensor. *Mater. Today Commun.* **2020**, 101077. [CrossRef]
36. Carlone, P.; Rubino, F.; Paradiso, V.; Tucci, F. Multi-scale modeling and online monitoring of resin flow through dual-scale textiles in liquid composite molding processes. *Int. J. Adv. Manuf. Technol.* **2018**, *96*, 2215–2230. [CrossRef]
37. Di Fratta, C.; Klunker, F.; Ermanni, P. A methodology for flow-front estimation in LCM processes based on pressure sensors. *Compos. Part A Appl. Sci. Manuf.* **2013**, *47*, 1–11. [CrossRef]
38. Kahali Moghaddam, M.; Salas, M.; Ersöz, I.; Michels, I.; Lang, W. Study of resin flow in carbon fiber reinforced polymer composites by means of pressure sensors. *J. Compos. Mater.* **2017**, *51*, 3585–3594. [CrossRef]
39. Liebers, N.; Bertling, D. Reducing NDT effort by coupled monitoring and simulation of liquid composite molding processes. In Proceedings of the 11th Symposium on NDT in Aerospace, Paris, France, 13–15 November 2019.
40. Konstantopoulos, S.; Fauster, E.; Schledjewski, R. Monitoring the production of FRP composites: A review of in-line sensing methods. *Express Polym. Lett.* **2014**, *8*. [CrossRef]
41. Gholizadeh, S. A review of non-destructive testing methods of composite materials. *Procedia Struct. Integr.* **2016**, *1*, 50–57. [CrossRef]

42. Collombet, F.; Torres, M.; Douchin, B.; Crouzeix, L.; Grunevald, Y.-H.; Lubin, J.; Camps, T.; Jacob, X.; Luyckx, G.; Wu, K.-T. Multi-instrumentation monitoring for the curing process of a composite structure. *Measurement* **2020**, *157*, 107635. [CrossRef]

43. El-Sabbagh, A.; Steuernagel, L.; Ziegmann, G. Characterisation of flax polypropylene composites using ultrasonic longitudinal sound wave technique. *Compos. Part B Eng.* **2013**, *45*, 1164–1172. [CrossRef]

44. Lionetto, F.; López-Muñoz, R.; Espinoza-González, C.; Mis-Fernández, R.; Rodríguez-Fernández, O.; Maffezzoli, A. A Study on Exfoliation of Expanded Graphite Stacks in Candelilla Wax. *Materials* **2019**, *12*, 2530. [CrossRef]

45. Hanse Wampo, F.L.; Lemanle Sanga, R.P.; Maréchal, P.; Ntamack, G.E. Piezocomposite transducer design and performance for high resolution ultrasound imaging transducers. *Int. J. Comput. Mater. Sci. Eng.* **2019**, *8*. [CrossRef]

46. Samet, N.; Marechal, P.; Duflo, H. Monitoring of an ascending air bubble in a viscous fluid/fiber matrix medium using a phased array transducer. *Eur. J. Mech. B/Fluids* **2015**, *54*, 45–52. [CrossRef]

47. Maréchal, P.; Ghodhbani, N.; Duflo, H. High temperature polymerization monitoring of an epoxy resin using ultrasound. In Proceedings of the 5th Global Conference on Polymer and Composite Materials (PCM 2018), Kitakyushu, Japan, 10–13 April 2018.

48. Dell'Anna, R.; Lionetto, F.; Montagna, F.; Maffezzoli, A. Lay-Up and Consolidation of a Composite Pipe by In Situ Ultrasonic Welding of a Thermoplastic Matrix Composite Tape. *Materials* **2018**, *11*, 786. [CrossRef] [PubMed]

49. Cascardi, A.; Dell'Anna, R.; Micelli, F.; Lionetto, F.; Aiello, M.A.; Maffezzoli, A. Reversible techniques for FRP-confinement of masonry columns. *Constr. Build. Mater.* **2019**, *225*, 415–428. [CrossRef]

50. Lionetto, F.; Morillas, M.N.; Pappadà, S.; Buccoliero, G.; Villegas, I.F.; Maffezzoli, A. Hybrid welding of carbon-fiber reinforced epoxy based composites. *Compos. Part A Appl. Sci. Manuf.* **2018**, *104*, 32–40. [CrossRef]

51. Liu, H.L.; Tu, X.-C.; Lee, J.O.; Kim, H.-B.; Hwang, W.R. Visualization of resin impregnation through opaque reinforcement textiles during the vacuum-assisted resin transfer molding process using ultrasound. *J. Compos. Mater.* **2014**, *48*, 1113–1120. [CrossRef]

52. Stöven, T.; Weyrauch, F.; Mitschang, P.; Neitzel, M. Continuous monitoring of three-dimensional resin flow through a fibre preform. *Compos. Part A Appl. Sci. Manuf.* **2003**, *34*, 475–480. [CrossRef]

53. Konstantopoulos, S.; Grössing, H.; Hergan, P.; Weninger, M.; Schledjewski, R. Determination of the unsaturated through-thickness permeability of fibrous preforms based on flow front detection by ultrasound. *Polym. Compos.* **2018**, *39*, 360–367. [CrossRef]

54. Zima, B.; Rafael, K. Numerical Study of Concrete Mesostructure Effect on Lamb Wave Propagation. *Material* **2020**, *13*, 2570. [CrossRef]

55. Jeong, J.J.; Choi, H. An impedance measurement system for piezoelectric array element transducers. *Measurement* **2017**, *97*, 138–144. [CrossRef]

56. Choi, H.; Jeong, J.J.; Kim, J. Development of an Estimation Instrument of Acoustic Lens Properties for Medical Ultrasound Transducers. *J. Healthc. Eng.* **2017**, *2017*. [CrossRef]

57. Lionetto, F.; Maffezzoli, A. Relaxations during the postcure of unsaturated polyester networks by ultrasonic wave propagation, dynamic mechanical analysis, and dielectric analysis. *J. Polym. Sci. Part B Polym. Phys.* **2005**, *43*, 596–602. [CrossRef]

58. Aziz, A.R.; Ali, M.A.; Zeng, X.; Umer, R.; Schubel, P.; Cantwell, W.J. Transverse permeability of dry fiber preforms manufactured by automated fiber placement. *Compos. Sci. Technol.* **2017**, *152*, 57–67. [CrossRef]

59. Lionetto, F.; Maffezzoli, A.; Ottenhof, M.A.; Farhat, I.A.; Mitchell, J.R. Ultrasonic investigation of wheat starch retrogradation. *J. Food Eng.* **2006**, *75*, 258–266. [CrossRef]

60. Lionetto, F.; Dell'Anna, R.; Moscatello, A.; Totaro, G.S.; Pappadà, A.M. Permeability of stitched and hot-bonded preforms for resin infusion. In Proceedings of the SAMPE Europe Conference, Southampton, UK, 15–17 September 2018.

61. Karaki, M.; Hallal, A.; Younes, R.; Trochu, F.; Lafon, P.; Hayek, A.; Kobeissy, A.; Fayad, A. A comparative analytical, numerical and experimental analysis of the microscopic permeability of fiber bundles in composite materials. *Int. J. Compos. Mater* **2017**, *7*, 82–102.

62. Gebart, B.R. Permeability of unidirectional reinforcements for RTM. *J. Compos. Mater.* **1992**, *26*, 1100–1133. [CrossRef]
63. Berdichevsky, A.L.; Cai, Z. Preform permeability predictions by self-consistent method and finite element simulation. *Polym. Compos.* **1993**, *14*, 132–143. [CrossRef]

Article

Numerical Study of Concrete Mesostructure Effect on Lamb Wave Propagation

Beata Zima * and Rafał Kędra

Department of Mechanics of Materials and Structures, Faculty of Civil and Environmental Engineering,
Gdańsk University of Technology, Narutowicza 11/12, 80-233 Gdansk, Poland; rafal.kedra@pg.edu.pl
* Correspondence: beata.zima@pg.edu.pl

Received: 6 May 2020; Accepted: 3 June 2020; Published: 4 June 2020

Abstract: The article presents the results of the numerical investigation of Lamb wave propagation in concrete plates while taking into account the complex concrete mesostructure. Several concrete models with randomly distributed aggregates were generated with the use of the Monte Carlo method. The influence of aggregate ratio and particle size on dispersion curves representing Lamb wave modes was analyzed. The results obtained for heterogeneous concrete models were compared with theoretical results for homogeneous concrete characterized by the averaged macroscopic material parameters. The analysis indicated that not only do the averaged material parameters influence the dispersion solution, but also the amount and size of aggregate particles. The study shows that Lamb waves propagate with different velocities in homogeneous and heterogeneous models and the difference increases with aggregate ratio and particle size, which is a particularly important observation for wave-based diagnostic methods devoted to concrete structures.

Keywords: concrete; mesostructure; Lamb wave; heterogeneity; Monte Carlo method; SHM

1. Introduction

Ultrasonic waves have been widely used in diagnostics of engineering structures. Significant scientists' interest has been focused on the nondestructive evaluation of concrete, which is one of the most popular construction materials used in the world. Its widespread use entails the need for the development of effective methods of damage detection in the early stage to prevent the development of significant defects, which would jeopardize the integrity and safety of the entire structure.

So far, a number of various approaches devoted to solving different problems have been proposed. Ultrasonic waves are most commonly used for crack detection [1–7], monitoring the hardening process [8–11], as well as for the quality assessment of the adhesive connection between concrete cover and reinforcement [12–16]. The method of scanning the surface opening cracks in reinforced concrete structures using transient elastic waves was developed by Liu et al. [1]. The effect of the depth of surface-breaking cracks in concrete plates on Lamb wave propagation was described by Yang et al. [2]. The influence of the width of partially closed surface-breaking cracks in concrete structures, the incident angle of waves with cracks, and the distance from the cracks on travel time and wave amplitude have been investigated by Pahlavan et al. [3]. Ultrasonic shear-wave tomography was used by Choi et al. [4] to identify horizontal cracks or delamination in concrete pavements, columns, and bridges. Surface wave propagation was used by Ham et al. [5] to assess the volume content of relatively small distributed defects and to characterize the microcrack networks in concrete. A combination of Rayleigh and longitudinal waves was employed by Aggelis and Shiotani [6] to evaluate the parameters of surface opening cracks in concrete before impregnation by the epoxy material, as well as to determine the final repair effectiveness. Quiviger et al. [7] conducted a simulation study of the influence of the depth and morphology of cracks in concrete on ultrasound diffusion.

Wave propagation was also effectively employed to monitor the concrete hardening process. The speed of propagation of ultrasound in various concrete samples varying in time of aging was investigated by del Rio et al. [8]. Piezoelectric transducers embedded in concrete structures were used by Dumoulin et al. [9] to monitor the setting and hardening phases of the early-age concrete. The hardening process of ultra-high-performance concrete was monitored by Lee et al. [10] using the characteristics of individual Lamb wave modes. The correlation between the shear wave velocity and penetration resistance for mortar mixtures was presented by Liu et al. [11].

Several scientific papers were devoted to guided waves for the detection of debonding occurring between concrete and internal or external reinforcement [12–15]. Monitoring of the interfacial debonding of a concrete-filled pultrusion-GFRP tubular column based on stress wave propagation was conducted by Yang et al. [12]. The wavelet packet-based energy index was proposed by Jiang et al. [13] to detect debonding between a steel beam and a carbon fiber-reinforced polymer plate. The time-reversal method was applied by Zhao et al. [14] to detect and localize the debonding along the steel–concrete interface. Giri et al. [15] employed the partial least-squares regression technique for detecting gaps in the steel–concrete composite specimens.

The above brief review of the literature indicates the multiplicity of investigation strands concerning wave-based concrete diagnostics carried out in the last decades. However, in the majority of reported cases, regardless of the considered problem, the diagnostic process requires some reference data collected for an intact structure, which would be compared to the data obtained for the damaged specimen. Due to obvious reasons, it is not always possible to make the baseline measurement for a pristine structure, and thus, the development of reference-free diagnostic procedures is of particular importance. The main idea of the recently developed reference-free methods is based on the comparison of the experimental measurement with the theoretical predictions [16]. The greater the difference between experimental and theoretical results, the greater the state of deterioration is expected. However, the discrepancies between experimental measurements and theoretical predictions were caused not only by the damage occurrence but also the simplification of the theoretical model of wave propagation in the concrete structure. The theoretical models were usually developed for homogeneous, isotropic materials characterized by the averaged macroscopic parameters, while the concrete is a multi-phase, strongly heterogeneous material. The complex mesostructure of the concrete led to inaccuracies in the wave propagation model. The theoretical predictions differed slightly from the actual results registered for experimental specimens, which, in consequence, led to inaccuracies in damage size assessment.

Although the papers published by Xu et al. [17], Abo-Quadis [18], and Ramaniraka et al. [19] deal with the impact of the concrete mesostructure on waves characteristics, the problem of wave propagation in heterogeneous materials is still not considered in detail, yet. The existence of aggregates varying in size, shape, and material parameters, and pores and cracks may lead to additional disturbing phenomena like scattering, diffractions, and wave attenuation affecting wave amplitude or propagation velocity, which were commonly used as indicative parameters in the above-mentioned studies. The wave propagation phenomenon may significantly differ for two different concrete specimens despite their identical macroscopic material parameters. Thus, the comprehensive analysis of wave propagation in concrete specimens taking into consideration their complex mesostructure is crucial for the further development of wave-based concrete diagnostic methods.

The main contribution of the paper is the analysis of the aggregate size and content on Lamb wave propagation. The algorithm of the generation of heterogeneous concrete models was developed and described step by step. Several numerical plate models differing in parameters of the internal mesostructure were developed using the Monte Carlo method. The averaged macroscopic parameters determined theoretically were compared to material parameters determined based on the shape of reconstructed dispersion curves for antisymmetric Lamb modes. The results obtained show that the concrete mesostructure has a significant influence on the wave propagation phenomenon. The wave velocity determined for the homogeneous concrete model differed from the velocity in the heterogeneous concrete model, even if the averaged material parameters were the same, and according to the theoretical wave propagation model, differences in velocities should not be observed. The difficulties in exact wave velocity determination must be taken into account, especially if the developed diagnostic procedure involves monitoring of wave propagation velocity.

2. Theoretical Background

2.1. Theory of Lamb Waves

The existence of Lamb waves was described by Horace Lamb in 1917 [20]. Lamb waves are guided between two parallel free surfaces and can thus be sustained in thin plates. Due to the dispersive nature of Lamb waves, their velocity and number of wave modes depend on excitation frequency. In general, two different mode families can be distinguished: Symmetric and antisymmetric wave modes. When the wave motion is associated with symmetrical displacement and stresses with respect to the middle plane, the symmetric wave mode propagates (Figure 1a). Antisymmetric displacements and stresses are associated with antisymmetric modes propagation (Figure 1b). The relationship between wavenumber and frequency for symmetric modes and antisymmetric modes can be obtained by solving the following dispersion equations, respectively:

$$\frac{\tan(qd)}{\tan(pd)} = -\frac{4k^2 pq}{\left(k^2 - q^2\right)^2} \tag{1}$$

$$\frac{\tan(qd)}{\tan(pd)} = -\frac{\left(k^2 - q^2\right)^2}{4k^2 pq} \tag{2}$$

where k is the wavenumber, d is specimen thickness, and p and q depend on angular frequency ω:

$$p^2 = \frac{\omega^2}{c_L^2} - k^2 \tag{3}$$

$$q^2 = \frac{\omega^2}{c_T^2} - k^2 \tag{4}$$

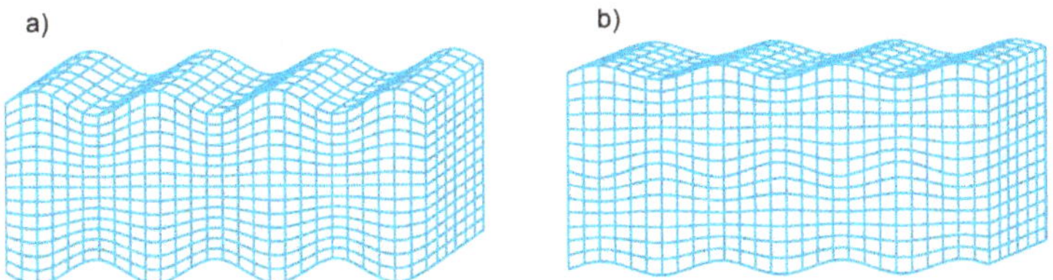

Figure 1. Lamb wave modes: (**a**) Symmetric mode; (**b**) antisymmetric mode.

The velocities of shear and longitudinal waves in an infinite medium denoted as c_T and c_L can be calculated based on known material parameters of the medium:

$$c_L = \sqrt{\frac{\lambda + 2\mu}{\rho}} \tag{5}$$

$$c_T = \sqrt{\frac{\mu}{\rho}} \tag{6}$$

where μ and λ are Lame's constants. The group velocity is the derivative of angular frequency over the wavenumber:

$$c_g = \frac{d\omega}{dk} \tag{7}$$

For the given angular frequency, there is an infinite number of possible solutions that fulfill Equations (1) and (2). The wavenumber k can be real, imaginary, or complex; however, if the plate is considered unloaded, it is sufficient to consider real values only. The number of possible solutions also increases with frequency range. For higher frequencies, the number of possible wave modes increases.

2.2. Two-Dimensional Fourier Transform

The shape of dispersion curves can be determined by solving the dispersion equations, but also by processing the signals captured at the investigated structure. It is possible to measure the amplitudes of individual Lamb wave modes by using a 2-dimensional fast Fourier transform (2D-FFT) technique [21]. In this approach, the time-domain propagation signals recorded at the series of equally spaced positions along the propagation path are transformed and the data from the time–space plane are converted into the frequency–wavenumber plane according to the following expression:

$$Y(k,\omega) = \int\limits_{-\infty}^{+\infty} \int\limits_{-\infty}^{+\infty} \mathbf{u}(x,t)e^{-i(kx+\omega t)}dxdt \tag{8}$$

where $\mathbf{u}(x,t)$ denotes the displacement of the surface.

3. Concrete Model Generation

3.1. Concrete Mesostructure

Concrete is a multi-phase, strongly heterogeneous material and its mesostructure presents strong randomness characteristics. It consists of aggregates varying in diameter and shape, mortar mix, interface transition zones (ITZs) located between mortar and aggregates, cracks, and pores. All concrete components are randomly distributed within its volume. The major part of the concrete mixture is aggregate particles. They can be divided into two groups by size: Fine ($\leq$4.75 mm) and coarse aggregate (>4.75 mm). The content of coarse aggregate in the concrete volume is usually equal to 40–50% and, thus, it largely determines the parameters of the concrete mixture, as well as its cost. The particle shape depends on the aggregate type. However, after other researchers [17,19,22] and for the sake of simplicity, in a later part of the paper, we assume that they are spherical.

The next component distinguished at the mesoscopic level is ITZs, which are the regions of the cement film covering the aggregate particles. As the first micro-cracks are induced at the interface, the ITZs are considered as the weak link in the concrete. Their negligible influence on guided wave propagation velocity was presented by Xu et al. [17] and, therefore, their existence is omitted in further investigations.

In this study, we consider a two-phase concrete model comprising mortar matrix and aggregate particles. In the following sections, the process of model generation is described step by step. The mesostructure of the concrete is generated using the Monte Carlo method, which allows for generation of random numbers with a specific distribution. The particles' diameters are chosen based on the a priori known grading curve, and particles' coordinates are generated by taking into account the certain restrictions related to the model dimensions.

3.2. Aggregate Particles Size

This section discussed the developed algorithm of diameter generation of the particles (Figure 2). The particle size distribution in concrete is expressed in terms of cumulative percentage passing through a series of sieves with openings with different sizes. The contribution of particle size fractions is usually presented in the form of a grading curve. The most popular and commonly used grading curve, which lies within the limiting grading curves proposed by design recommendations [23], is the empirical curve proposed by Fuller [22,24]. Fuller's curve provides optimum density and strength of the concrete and is described by the following formula:

$$P(d) = \left(\frac{d}{d_{max}}\right)^n \cdot 100\% \tag{9}$$

where $P(d)$ is the cumulative percentage passing through the sieve with diameter opening d, d_{max} is the maximum particle size, and n is the exponential factor with a typical value of 0.4–0.7. The weight percentage of the individual grading segment can be calculated as

$$P[d_s, d_{s+1}] = P(d_s) - P(d_{s+1}) \tag{10}$$

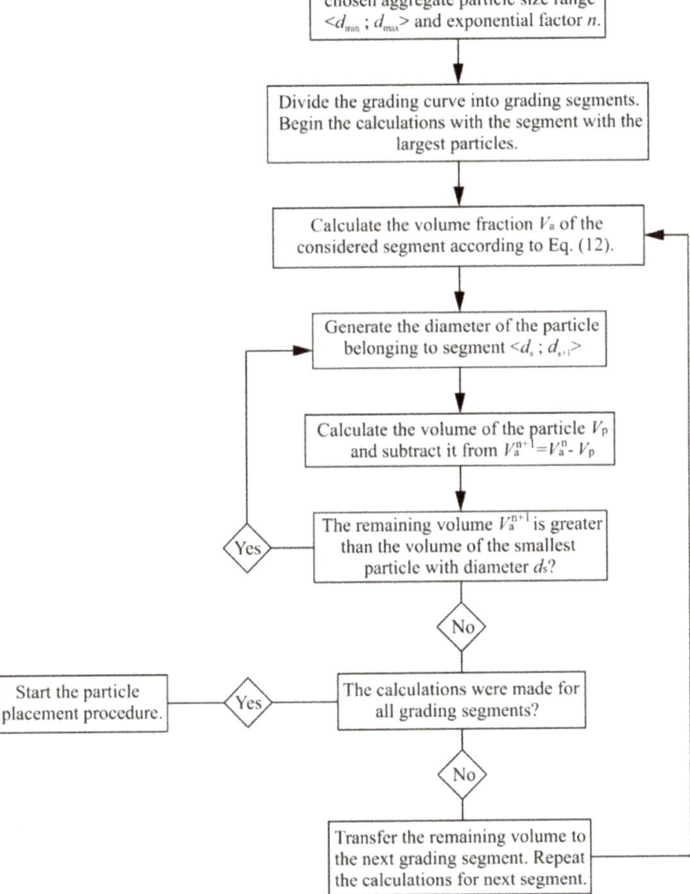

Figure 2. Generation of the aggregate particles.

The aggregate volume fraction in the concrete mixture is

$$v_a = \frac{m_a}{\rho_a V} \tag{11}$$

where m_a is the mass of aggregate particles, ρ_a is aggregate material density, and V is the volume of the concrete specimen. The volume of the particles in the grading segment $V_a[d_s, d_{s+1}]$ can then be calculated as

$$V_a[d_s, d_{s+1}] = \frac{P(d_s) - P(d_{s+1})}{P(d_{max}) - P(d_{min})} \cdot v_a \cdot V \tag{12}$$

To generate the particle belonging to the grading segment with randomly chosen diameter, one can use the formula:

$$d_p = \eta(d_{s+1} - d_s) + d_s \tag{13}$$

where η is a uniformly distributed random number in the interval (0,1). The number of particles generated in this way must ensure that the difference between the sum of their volumes and the volume $V_a[d_s, d_{s+1}]$ is smaller than the volume of the smallest particle vs. belonging to the considered segment:

$$V_a[d_s, d_{s+1}] - \sum_i^n V_p^i > V_s \tag{14}$$

Based on the described relationships, the algorithm of diameter generation of the particles was performed.

3.3. Particle Placement

Next, the generated particles must be placed into the concrete volume. As mentioned, they are randomly distributed; however, their placement must satisfy some primary conditions. First, all particles must be located within the concrete volume, and secondly, none of them can overlap with each other. Moreover, the placement process should include that the distance $s_{i,j}$ between the centers of adjacent particles must be greater than the sum of their radii because of the mortar film covering the aggregates (Figure 3):

$$s_{i,j} \geq \frac{d_i + d_j}{2} + \gamma_{i,j} \tag{15}$$

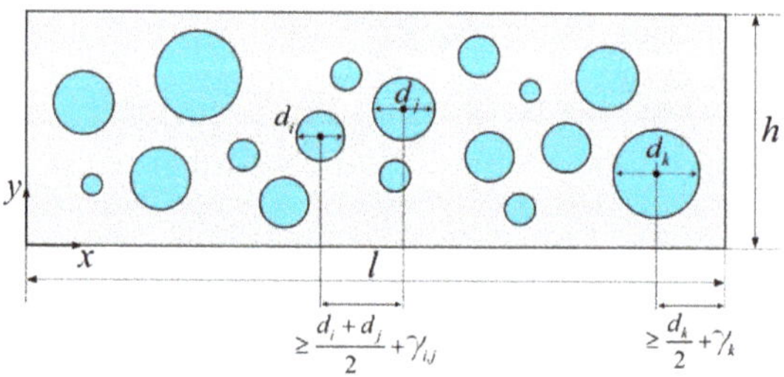

Figure 3. Particle placement in concrete specimen.

In this study, the thickness of mortar film is equal to 5% of the sum of the diameters of adjacent particles [24]:

$$\gamma_{i,j} = 0.05 \cdot (d_i + d_j) \tag{16}$$

Additionally, the distance between the particle and the specimen boundary must be at least equal to:

$$\gamma_i = 0.05 \cdot d_i \tag{17}$$

The above conditions imply that the coordinates of the mass center of the particle with diameter d located within the volume of the concrete plate with thickness h and length l must be generated in the following way:

$$x_i = \eta(x_{\max} - x_{\min}) + x_{\min} = \eta(l - 1.1d_i) + 0.55d_i \tag{18}$$

$$y_i = \eta(y_{\max} - y_{\min}) + y_{\min} = \eta(h - 1.1d_i) + 0.55d_i \tag{19}$$

where η is the uniformly distributed random number in the interval (0,1).

The algorithm of the particle placement can be summarized in three steps:

- Step I: For the particle with diameter d_i, generate the coordinates of its mass center. It is recommended to begin from the largest particles, which significantly facilities the placement process.
- Step II: Check if the particle covered with the mortar film does not overlap with any other previously generated particle. If it does, generate new coordinates and check again. If particles do not overlap, generate the coordinates for the next particle.
- Step III: Repeat Steps I and II for all particles.

3.4. Description of Numerical Models

3.4.1. Model Geometry and Material Parameters

The plate model developed in the Abaqus/Explicit environment has the dimensions 50 mm × 500 mm (Figure 4). The adopted material parameters of the mortar matrix and aggregate are summarized in Table 1 [17,25]. The calculations were performed for nine models varying in aggregate ratio and aggregate particle size (Table 2). The grading curves for particular models are presented in Figure 5. In all cases, the exponential factor n was equal to 0.5. Particles smaller than 4 mm in diameter were not introduced in any model, to avoid the problems with mesh generation.

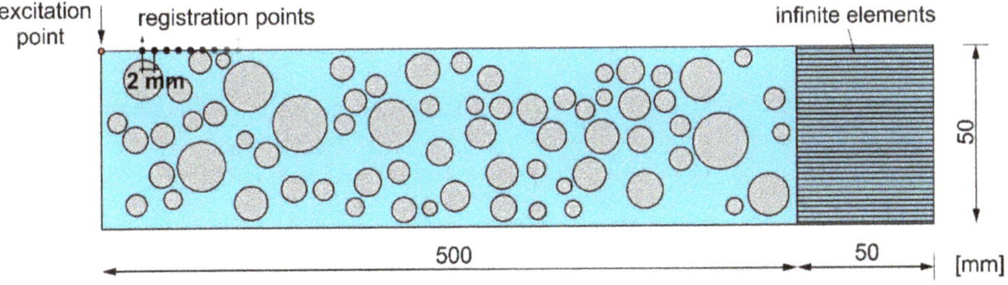

Figure 4. Geometry of numerical models.

Table 1. Material parameters of matric mortar and aggregate particles.

Material Parameter	Mortar Matrix	Aggregate Particles
Elastic modulus (GPa)	26	60
Poisson's ratio (-)	0.2	0.22
Density (kg/m^3)	2100	2700

The models were developed with the use of four-node plane strain elements with reduced integration (CPE4R). To ensure that the measured signals are dominantly influenced by the concrete mesostructure, a layer of four-node infinite plane strain elements (CINPE4) with a thickness of 50 mm was additionally introduced. The layer of infinite elements allows us to avoid the registering of reflection from the edge of the plate. Moreover, they allow the size of the investigated structure to be reduced. The dimensions of the finite elements were 1 mm × 1 mm and the length of the integration step was to $\Delta t = 10^{-8}$ s. The dimensions of the infinite elements were 1 mm × 50 mm. To visualize the differences in aggregate particle concentration, the geometries of three exemplary models are presented in Figure 6. Additionally, Figure 7 shows histograms of the special distribution of the generated particles for three different aggregate ratios. It can be seen that the distribution is more or less uniform, which confirms the random generation of particle placement.

Table 2. Parameters of numerical models.

Model	Aggregate Ratio	Aggregate Particle Size (mm)
A1	20%	
A2	30%	4–12
A3	40%	
B1	20%	
B2	30%	6–14
B3	40%	
C1	20%	
C2	30%	8–16
C3	40%	

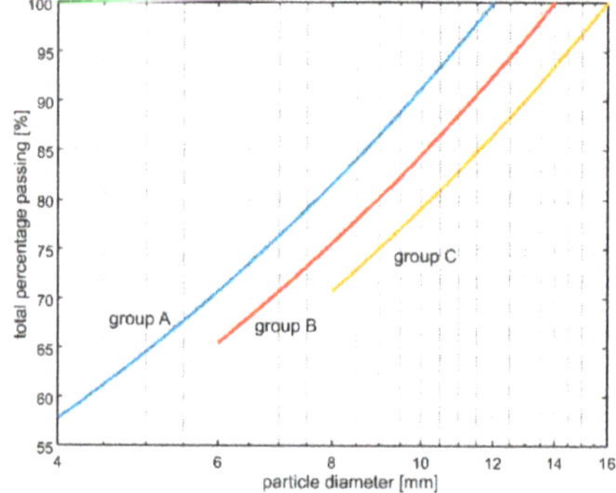

Figure 5. Fuller's curves for concrete models considered in the study.

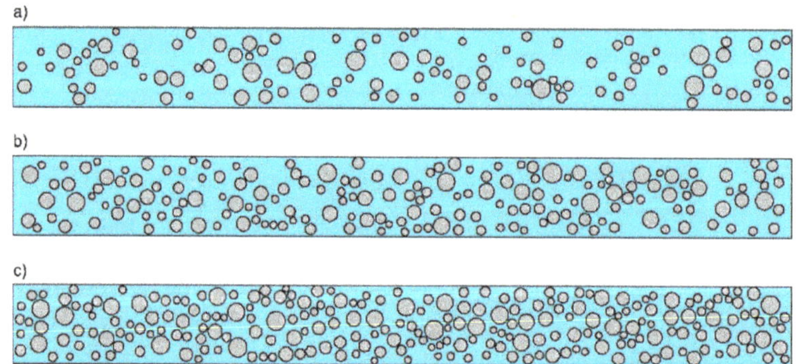

Figure 6. Geometry of numerical models varying in aggregate ratio: (**a**) Model A1, (**b**) model A2, and (**c**) model A3.

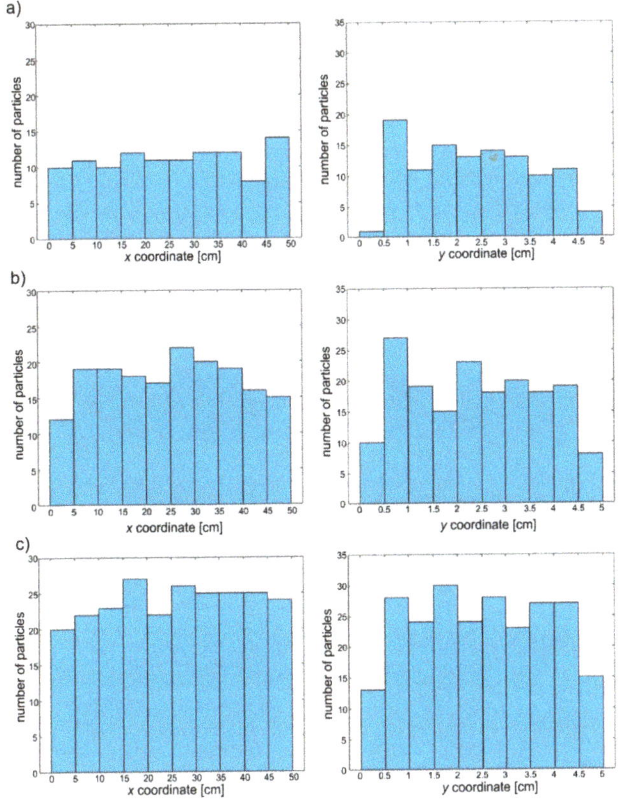

Figure 7. Histograms of the special distribution of the generated particles for model (**a**) A1, (**b**) A2, and (**c**) A3.

3.4.2. Excitation Function

The excitation function was in the form of a windowed tone burst, which is commonly used in Lamb wave-based inspections [26]. The tone burst is a modulated sine or cosine function by the Hann or Gaussian window. The modulation allows the dispersion effect to be reduced and provides the mode purity [26]. In this study, the five-cycle sine function with carrier frequencies of 25, 50, 100, and 150 kHz modulated by the Hann window was used. In the numerical model, the excitation was applied as a concentrated time-dependent force. As, in the actual cases, the actuators and sensors are usually attached at the surface of the plate, the excitation was applied perpendicularly to the plate surface. The signals were registered at the series of positions spaced 1 mm apart along the propagation path with a length of 49.5 cm.

4. Results

4.1. Visualization of Lamb Wave Propagation

Visualization of the propagating wave in the form of magnitudes of displacements at selected time instants is presented in Figure 8. For comparison, the figure also contains the results for a homogeneous plate characterized by mortar parameters (aggregate ratio 0%). It can be seen that in the case of the homogeneous plate, the displacements of outer surfaces are the same and the map is symmetrical with respect to the middle plane, which clearly indicates the existence of an antisymmetric mode. The presence of an even smaller number of aggregate particles significantly affects the observed wave motion and mode purity. The displacements of the outer surfaces in model A1 are not the same. Moreover, the symmetry of the displacement map is not sustained (Figure 8b). The symmetry disruption becomes more visible in the model with a higher aggregate ratio (Figure 8c). In the case of heterogeneous models, the propagating wave reflects from the particles characterized by different material parameters than the mortar matrix. Part of the reflected wave energy propagates back along the plate model. Note that, in the case of the homogeneous model after 0.22 ms, wave motion is not observed at the initial part of the model, while in heterogeneous models, low-amplitude wave motion resulting from wave interactions with particles is observed in the entire plate volume. It is noteworthy that additional wave phenomena like diffractions and reflections resulting in additional peaks registered in signals may lead to significant difficulties in interpretation of the results.

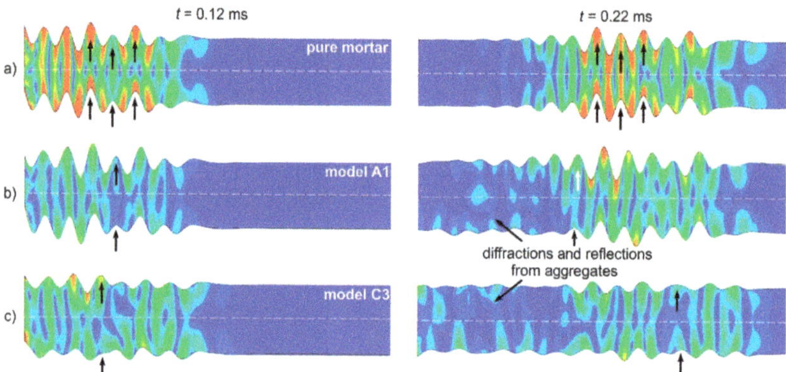

Figure 8. Visualization of wave propagation in (**a**) homogeneous concrete plate, (**b**) heterogeneous model A1, and (**c**) heterogeneous concrete model C3.

The aggregate ratio also affected the dissipated energy. The wave amplitude is clearly higher for the model made of pure mortar. The displacements caused by wave motion becomes lower for increasing aggregate ratio.

4.2. Dispersion Curves

Figures 9 and 10 present the 2-D view of the wavenumber–frequency information, which was obtained by carrying out the 2D-FFT (see Equation (8)) of the time series registered for four different excitation frequencies in equally spaced points on the plate surface. The final map presents the dispersion curves representing antisymmetric Lamb wave modes described by Equation (2). Based on the 2D-FFT results, the shape of the curve for the first antisymmetric mode was reconstructed. For comparison, the wavenumber–frequency information obtained for the homogeneous concrete model is presented in Figure 9, while Figure 10 contains the results for nine heterogeneous models.

As we can see, the aggregate ratio clearly influences the visibility of the particular dispersion curves. In the case of the homogeneous material, the curves can be unambiguously distinguished. The presence of aggregate particles and aggregate-wave interactions affecting the signals' characteristics resulted in deterioration of the quality of the maps. The map with the worst quality was obtained for model C3 characterized by the highest aggregate ratio and the largest particles (see Table 2). Heterogeneity hinders the reconstruction of dispersion curves based on wavenumber–frequency information.

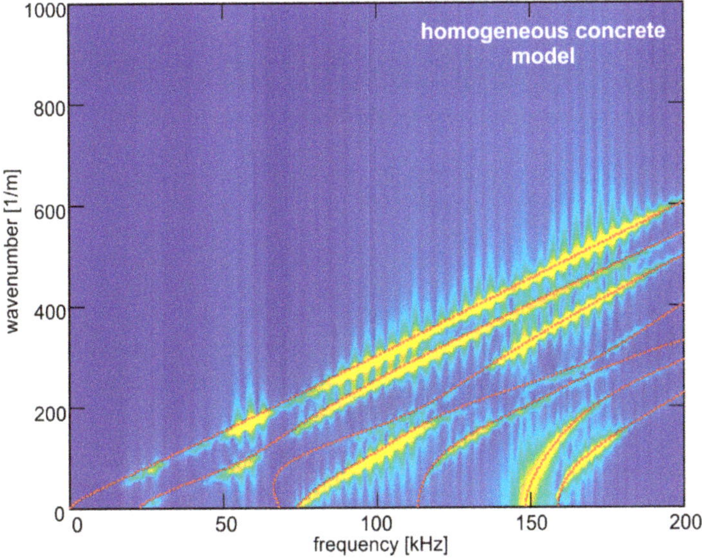

Figure 9. Wavenumber–frequency representation for homogeneous concrete model.

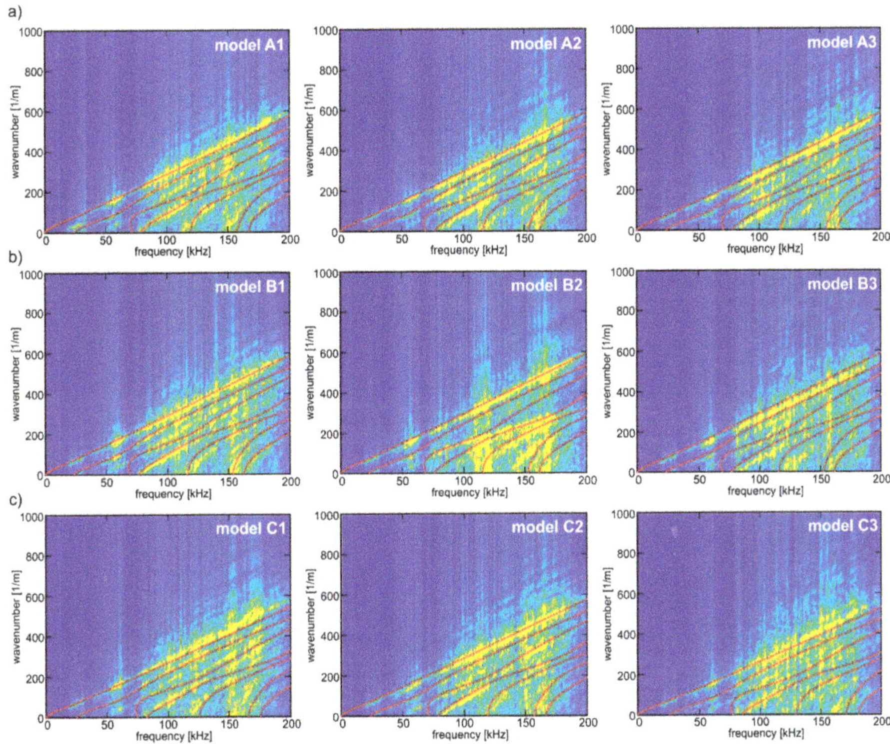

Figure 10. Wavenumber–frequency maps determined for numerical models of heterogeneous concrete plates: (**a**) group A; (**b**) group B and (**c**) group C.

4.3. Determining the Elastic Modulus of Concrete Plates

To compare the numerical results with theoretical predictions, the elastic modulus was estimated in two ways. First, the shape of dispersion curves was reconstructed based on numerical results. Secondly, it was calculated based on known proportions between aggregate particles and mortar matrix.

The elastic modulus identification procedure was developed in a MATLAB environment using function fminsearch. The procedure of curve shape reconstruction can be summarized in four steps. In the first step, the data obtained from the FEM analysis were transformed in the wavenumber–frequency domain using a 2D-FFT-based algorithm. Further considerations were conducted for the limited representative area of data (Figure 10a). In our case, due to the use of excitation in the form of wave packets with central frequencies of 25, 50, 100, and 150 kHz, the frequency range, which was taken into account, was 0–250 kHz and the corresponding wavenumber range was 0–1000 m^{-1}. To optimize the calculation process, the 2D-FFT results were normalized with respect to the maximum amplitude value for each frequency:

$$\hat{Y}(k_i, \omega_i) = \frac{Y(k_i, \omega_i)}{\max\{Y(\omega_i)\}} \tag{20}$$

The normalization process excluded the unequal influence of the individual frequencies. Moreover, there was no need to use weight functions. Examples of normalized data are shown in Figure 11b. Normalization was the last stage of data preparation. In the next step, the procedure for determining the elastic modulus was initiated. In the first step, the dependence for the first antisymmetric Lamb mode was calculated for the pre-established Young modulus. The determined values creating a dispersion curve were considered as the sets of n pairs of numbers:

$$K = \{(\Delta\omega, k(\Delta\omega)), (2\Delta\omega, k(2\Delta\omega)), \dots, (n\Delta\omega, k(n\Delta\omega))\} = \{(\Delta\omega, k_1), (2\Delta\omega, k_2), \dots, (n\Delta\omega, k_n)\} \tag{21}$$

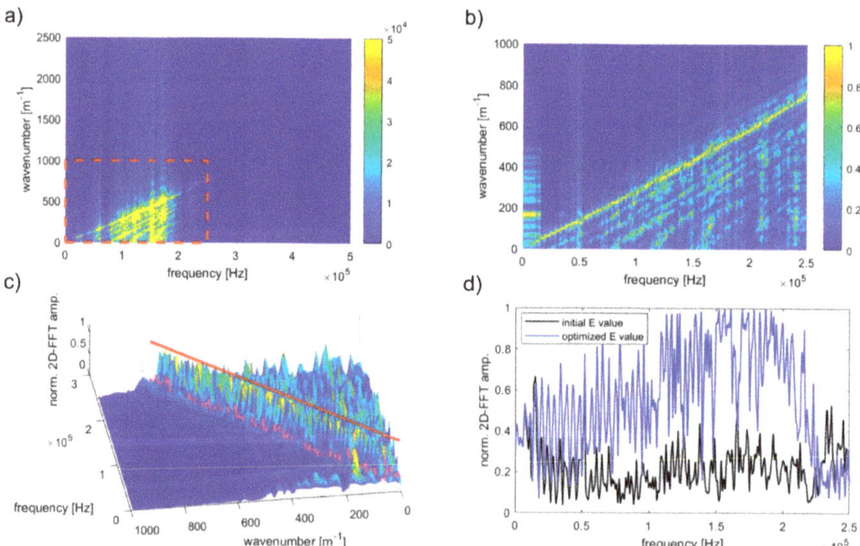

Figure 11. The Young modulus determining algorithm: (**a**) The 2D-FFT map and the data area selected for further analysis; (**b**) results of the normalization; (**c**) visualization of 2D-FFT interpolation along the dispersion curve; (**d**) comparison of interpolation results for a pre-selected and finally appointed Young modulus values.

Next, the standardized 2D-FFT values for particular angular frequencies were interpolated, forming a second set of data (Figure 10c,d):

$$\hat{Y}_{\text{int}} = \{\hat{Y}(\Delta\omega, k_1), \hat{Y}(2\Delta\omega, k_2), \ldots, \hat{Y}(n\Delta\omega, k_n)\} = \{\hat{Y}_1, \hat{Y}_2, \ldots, \hat{Y}_{n,}\} \tag{22}$$

Finally, the value of the following function was calculated:

$$\hat{F} = \frac{1}{n\sum\limits_{i}^{n} \hat{Y}_i^2} \tag{23}$$

The procedure was repeated for different values of elastic moduli until the function $\hat{F}$ reaches a minimum value. The minimum value of the function $\hat{F}$ indicated that the dispersion curve that analytically determined the best coincides with the dispersion curve visible in the map. This procedure was tested first for the homogeneous material with parameters $E = 26$ GPa, $v = 0.2$, and $\rho = 2100$ kg/m^3. The elastic modulus determined based on numerical results was 25.92 GPa and the percentage error was 0.309%, which is a satisfactory consistency of results. The analytical dispersion curves for the finally determined Young modulus were imposed on the numerical maps by red dashed lines (Figure 10), while the corresponding values of Young's modulus denoted as E_c^{DC} are summarized in Table 3.

Table 3. Parameters of numerical models.

Model	Aggregate Volume Fraction (-)	Density (kg/m^3)	E_c^V acc. Voigt Model (GPa)	E_c^R acc. Reuss Model (GPa)	E_c^{DC} Based on Dispersion Curves (GPa)
homogeneous concrete model	0	2100	26	26	25.92
A1	0.163	2197.8	31.542	28.650	30.310
A2	0.250	2250.0	34.500	30.290	29.459
A3	0.342	2305.2	37.628	32.250	35.290
B1	0.163	2197.8	31.542	28.650	28.734
B2	0.250	2250.0	34.500	30.290	29.119
B3	0.342	2305.2	37.628	32.250	35.689
C1	0.163	2197.9	31.542	28.650	29.400
C2	0.250	2250.0	34.500	30.290	29.104
C3	0.342	2305.2	37.628	32.250	37.475

The second stage of the analysis involved the theoretical calculation of Young's modulus. There are several theoretical models, which allow the elastic modulus of two-phase heterogeneous materials to be estimated. Two most common approaches were proposed by Voigt and Reuss. The use of either of these two models requires knowledge of the modulus of elasticity of mortar and aggregate and the volume of aggregates. According to the Reuss model, the elastic modulus is calculated in the following way:

$$E_C^R = \frac{E_m}{1 + \left(\frac{E_m}{E_a} - 1\right)V_a} \tag{24}$$

The overall elastic modulus of concrete by Voigt is:

$$E_C^V = E_m\left(1 + \left(\frac{E_a}{E_m} - 1\right)V_a\right) \tag{25}$$

where E_m and E_a are elastic moduli of mortar and aggregate, respectively, and V_a is the volume fraction of aggregate. The elastic modulus calculated for all nine models is summarized in Table 3. Additionally, the result for the homogeneous concrete model was added for comparison. It can be seen that if $E_a > E_m$, the Voigt model always predicts higher values of the elastic modulus than the Reuss model. The experimental study presented in previously published papers showed that these two models usually define the upper and lower bound of the concrete elastic modulus and the exact value usually lies between their predictions [27]. Indeed, the modulus value predicted on the basis of dispersion curves E_c^{DC} lies between theoretically determined boundary values. The differences between particular results, as well as the percentage errors, are reported in Table 4.

Table 4. Differences between theoretical and numerical results.

Model	The Difference $\|E_c^V - E_c^{DC}\|$ (GPa)	Percentage Error $\frac{\|E_c^V - E_c^{DC}\| \cdot 100\%}{E_c^{DC}}$ (%)	The Difference $\|E_c^R - E_c^{DC}\|$ (GPa)	Percentage Error $\frac{\|E_c^R - E_c^{DC}\| \cdot 100\%}{E_c^{DC}}$ (%)
homogeneous concrete model	0.08	0.309	0.08	0.309
A1	1.232	4.065	1.66	5.477
A2	5.041	17.111	0.831	2.820
A3	2.338	6.625	3.040	8.614
B1	2.808	9.772	0.084	0.292
B2	5.381	18.479	1.171	4.021
B3	1.939	5.433	3.439	9.636
C1	1.842	6.265	0.750	2.551
C2	5.396	18.540	1.186	4.075
C3	0.153	0.408	5.225	13.943

It can be seen that Young's modulus calculated according to both theoretical models clearly increased with the volume fraction of aggregate particles. The results obtained using the dispersion curves do not show the same increasing tendency. As expected, the elastic modulus was found to be the highest for models A3, B3, and C3, but the value of E_c^{DC} was higher for model A1 than for A2, and that for model C1 was higher than that for C2. This means that the aggregate presence affected wave propagation velocity. Moreover, the wave-aggregate interactions could affect the signal characteristics, which, in turn, resulted in a change in the shape of the dispersion curves.

Comparing the values of percentage errors reported in Table 4, it is clearly visible that the theoretical Reuss models are better suited to numerical results obtained by dispersion curves' reconstruction. The average percentage error for the Reuss model is 5.714%, while for the Voigt model, it is 9.633%. The greatest differences were reported for models with an aggregate ratio of 40%: 8.614% for A3, 9.636% for B3, and 13.943% for C3. One can conclude that the discrepancy between theoretical and experimental results clearly increases with the number of scatterers, but also with their size.

The presented results indicate that the heterogeneity of concrete influences wave propagation characteristics. The elastic modulus estimated using most common theoretical models differ from the modulus estimated based on dispersion curves. Meanwhile, the difference in modulus values leads to discrepancies in wave velocity estimation, which is particularly important if the wave velocity is used as an indicative parameter in the diagnostic process. To illustrate the differences in wave velocity, the Lamb dispersion equations were solved for model C3, for which the highest error was noted. Figure 12 contains a comparison of the first symmetric and antisymmetric modes. Disregarding the impact of the concrete mesostructure may lead to incorrect velocity determination—for some frequencies, the discrepancy may reach over 400 m/s.

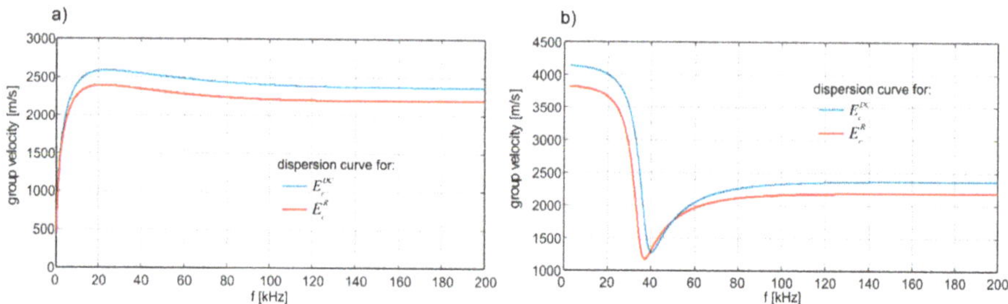

Figure 12. Comparison of (**a**) antisymmetric and (**b**) symmetric modes traced for various values of Young's modulus.

5. Conclusions

The paper presents the results of the investigation of the influence of a concrete mesostructure on Lamb wave propagation. The numerical calculations of guided wave propagation were performed for two-phase concrete plate models varying in aggregate ratio and particle diameter size. The random distribution of aggregate particles was generated using the Monte Carlo method. The diameter size was generated according to Fuller's grading curve.

The numerical visualization showed that the heterogeneous mesostructure of the concrete plate affected the displacements associated with wave motion. The displacement maps were symmetric with respect to the middle plane only for a homogeneous concrete plate. The presence of even a small number of particles disrupted the displacement symmetry, as well as the wave amplitude.

The study also involved the comparison of Young's modulus determined in two different ways: Theoretically based on two models commonly used for two-phase heterogeneous materials (Reuss and Voigt model) and numerically by reconstructing the dispersion curves calculated with the use of time-domain signals registered at the plate surface after wave excitation. The values of elastic modulus obtained theoretically and numerically differ, and the discrepancy increased with aggregate ratio and the size of aggregate particles. The results obtained clearly indicate that a Lamb wave propagating in an isotropic, homogeneous material characterized by the averaged macroscopic parameters is characterized by different velocities than the wave propagating in a heterogeneous material. The discrepancy increased with an increasing number of scatterers associated with more often wave-aggregate interactions, affecting wave propagation. The influence of wave-aggregate interactions also increased with particle size as the diameter approached the wavelength. These observations are particularly important for diagnostic procedures, which use wave propagation velocity as an indicative parameter for structural state assessment.

Author Contributions: Conceptualization, B.Z.; methodology, B.Z. and R.K.; software, B.Z.; validation, B.Z. and R.K.; formal analysis, B.Z.; investigation, B.Z.; data curation, B.Z.; writing—original draft preparation, B.Z.; writing—review and editing, B.Z. and R.K.; visualization, B.Z. and R.K. All authors have read and agreed to the published version of the manuscript.

Funding: This research was funded by the Faculty of Civil and Environmental Engineering at Gdańsk University of Technology.

Conflicts of Interest: The authors declare no conflict of interest.

References

1. Liu, P.-L.; Lee, K.-H.; Wu, T.T.; Kuo, M.-K. Scan of surface-opening cracks in reinforced concrete using transient elastic waves. *NDT E Int.* **2001**, *34*, 219–226. [CrossRef]

2. Yang, Y.; Cascante, G.; Polak, M.-A. Depth detection of surface-breaking cracks in concrete plates using fundamental Lamb modes. *NDT E Int.* **2009**, *42*, 501–512. [CrossRef]

3. Pahlavan, L.; Zhang, F.; Blacquiere, G.; Yang, Y.; Hordijk, D. Interaction of ultrasonic waves with partially-closed cracks in concrete structures. *Constr. Build. Mater.* **2018**, *167*, 899–906. [CrossRef]

4. Choi, P.; Kim, D.-H.; Lee, B.-H.; Won, M.C. Application of ultrasonic shear-wave tomography to identify horizontal crack or delamination in concrete pavement and bridge. *Constr. Build. Mater.* **2016**, *12*, 81–91. [CrossRef]

5. Ham, S.; Song, H.; Oelze, M.; Popovics, J.S. A contactless ultrasonic surface wave approach to characterize distributed cracking damage in concrete. *Ultrasonics* **2017**, *75*, 46–57. [CrossRef]

6. Aggelis, D.G.; Shiotani, T. Repair evaluation of concrete cracks using surface and through-transmission wave measurement. *Cem. Concr. Compos.* **2007**, *29*, 700–711. [CrossRef]

7. Quiviger, A.; Girard, A.; Payan, C.; Chaix, J.F.; Garnier, V.; Salin, J. Influence of the depth and morphology of real cracks on diffuse ultrasound in concrete: A simulation study. *NDT E Int.* **2013**, *60*, 11–16. [CrossRef]

8. Del Rio, L.M.; Jimenez, A.; Lopez, F.; Rosa, F.J.; Rufo, M.M.; Paniagua, J.M. Characterization and hardening of concrete with ultrasonic testing. *Ultrasonics* **2004**, *42*, 527–530. [CrossRef]

9. Dumoulin, C.; Karaiskos, G.; Carette, J.; Staquet, S.; Deraemaeker, A. Monitoring of the ultrasonic P-wave velocity in early-age concrete with embedded piezoelectric transducers. *Smart Mater. Struct.* **2012**, *21*, 1–4. [CrossRef]

10. Lee, C.; Park, S.; Bolander, J.; Pyo, S. Monitoring the hardening process of ultra high performance concrete using decomposed modes of guided waves. *Constr. Build. Mater.* **2018**, *163*, 267–276. [CrossRef]

11. Liu, S.; Zhu, J.; Seeaj, S.; Cano, R.; Juenger, M. Monitoring setting and hardening process of mortar and concrete using ultrasonic shear waves. *Constr. Build. Mater.* **2014**, *72*, 248–255. [CrossRef]

12. Yang, W.; Yang, X.; Li, S. Monitoring of Interfacial Debonding of Concrete Filled Pultrusion-GFRP Tubular Column Based on Piezoelectric Smart Aggregate and Wavelet Analysis. *Sensors* **2020**, *20*, 2149. [CrossRef] [PubMed]

13. Jiang, J.; Jiang, J.; Deng, X.; Deng, Z. Detecting Debonding between Steel Beam and Reinforcing CFRP Plate Using Active Sensing with Removable PZT-Based Transducers. *Sensors* **2020**, *20*, 41. [CrossRef] [PubMed]

14. Zhao, G.; Zhang, D.; Zhang, L.; Wang, B. Detection of defects in reinforced concrete structures using ultrasonic nondestructive evaluation with piezoceramic transducers and time reversal method. *Sensors* **2018**, *18*, 4176. [CrossRef]

15. Giri, P.; Mishra, S.; Clark, S.M.; Samali, B. Detection of gaps in concrete-metal composite structures based on the feature extraction method using piezoelectric transducers. *Sensors* **2019**, *19*, 1769. [CrossRef]

16. Zima, B.; Kędra, R. Debonding size estimation in reinforced concrete beams using guided wave-based method. *Sensors* **2020**, *20*, 389. [CrossRef]

17. Xu, B.; Chen, H.; Mo, Y.-L.; Zhou, T. Dominance of debonding defect of CFST on PZT sensor response considering the meso-scale structure of concrete with multi-scale simulation. *Mech. Syst. Sig. Proc.* **2018**, *107*, 515–528. [CrossRef]

18. Abo-Qudais, S.A. Effect of concrete mixing parameters on propagation of ultrasonic waves. *Constr. Build. Mater.* **2005**, *19*, 257–263. [CrossRef]

19. Ramaniraka, M.; Rokotonarivo, S.; Payan, C.; Garnier, V. Effect of the Interfacial Transition Zone on ultrasonic wave attenuation and velocity in concrete. *Cem. Concr. Res.* **2019**, *124*, 105809. [CrossRef]

20. Lamb, H. On waves in elastic plate. *Proc. R. Soc. Lond. Ser. A* **1917**, *93*, 114–128.

21. Alleyne, D.; Cawley, P. A two-dimensional Fourier transform method for the measurement of propagating multimode signals. *J. Acoust. Soc. Am.* **1991**, *89*, 1159–1168. [CrossRef]

22. Wriggers, P.; Moftah, S.O. Mesoscale models for concrete: Homogenisation and damage behavior. *Finite Elem. Anal. Des.* **2006**, *42*, 623–636. [CrossRef]

23. *Concrete, Reinforced and Prestressed Concrete Structures*; German Institute for Standardisation: Berlin, Germany, 2008.

24. Schlangen, E.; van Mier, J.G.M. Simple lattice model for numerical simulation of fracture of concrete materials and structures. *Mater. Struct.* **1992**, *25*, 534–542. [CrossRef]

25. Du, X.; Jin, L.; Ma, G. Numerical modeling tensile failure behavior of concrete at mesoscale using the extended finite element method. *Int. J. Dam. Mech.* **2014**, *23*, 872–898. [CrossRef]

26. Giurgiutiu, V. *Structural Health Monitoring with Piezoelectric Wafer Active Sensors*; Academic Press: Cambridge, MA, USA, 2008.

27. Zhou, F.P.; Lydon, F.D.; Barr, B.I.G. Effect of coarse aggregate on elastic modulus and compressive strength of high performance concrete. *Cem. Concr. Res.* **1995**, *25*, 177–186. [CrossRef]

 *materials*

Article

Monitoring the Setting Process of Cementitious Materials Using Guided Waves in Thin Rods

Dongquan Wang [1], Guangyun Yu [2,*], Shukui Liu [1,2] and Ping Sheng [2]

[1] State Key Laboratory for Geomechanics and Deep Underground Engineering, China University of Mining and Technology, Xuzhou 221116, China; wdq8785@cumt.edu.cn (D.W.); skliu@cumt.edu.cn (S.L.)
[2] School of Mechanics and Civil Engineering, China University of Mining and Technology, Xuzhou 221116, China; shengping@cumt.edu.cn
* Correspondence: gyyu_cumt@163.com

Abstract: Characterizing early-age properties is very important for the quality control and durability of cementitious materials. In this paper, an approach using embedded guided waves was adopted to monitor the changes in the mechanical proprieties of mortar and concrete during setting, and embedded thin rods with low-cost piezoelectric sensors mounted on top were used for guide wave monitoring. Through continuous attenuation monitoring of the guided waves, the evolution of mortar and concrete properties was characterized. Four different kinds of metallic rods were tested at the same time to find out the optimal setup. Meanwhile, shear wave velocities of the mortar and concrete samples were monitored and correlated to the attenuation, and setting time tests were also performed on these samples. Experimental results demonstrate that the proposed approach could monitor the evolution of the setting of cementitious materials quantitatively, and time of the initial setting could be determined by this technique as well. In addition, it is found that the attenuations of fundamental longitudinal guided wave mode are almost the same in concrete samples and mortar samples sieved from concrete, indicating that this technique is able to eliminate the effects of coarse aggregates, which makes it of great potential for in-situ monitoring of early age concrete.

Keywords: guided waves; setting time; mortar and concrete; early age

Citation: Wang, D.; Yu, G.; Liu, S.; Sheng, P. Monitoring the Setting Process of Cementitious Materials Using Guided Waves in Thin Rods. *Materials* **2021**, *14*, 566. https://doi.org/10.3390/ma14030566

Academic Editors: Mathieu Bauchy, Francesca Lionetto and Sanjay Mathur
Received: 1 December 2020
Accepted: 21 January 2021
Published: 25 January 2021

1. Introduction

In reinforced concrete structures, the workability and durability of concrete are critical for the structures to meet desired structural performance [1–3]. Furthermore, the evaluation of properties, such as the setting time of fresh concrete, is critical for assuring quality and reducing construction time. Conventionally, the penetration test that measures the shear resistance of the cementitious material was widely used to determine setting times [4]. In addition, research also showed the possibility of using ultrasonic body waves (longitudinal and shear waves) to characterize the early age properties of cementitious material [5–14].

Lots of research has been conducted to correlate the body wave velocities and material properties. For instance, many efforts were made, aiming to find the relationship between primary wave velocity and setting times [9,11,15–19]. Dumoulin et al. [19] used embedded piezoelectric patches as smart aggregates, to monitor the primary wave velocity evolution during the setting and hardening phases of concrete, and it was found that it was hard to extract the primary wave velocity with a good accuracy at very early ages. Besides, water in the cementitious materials leads to a high primary wave velocity at the early age, which will shield the velocity originating from the solid portion of the material during setting [20]. Moreover, recent research showed that the primary wave velocity in cementitious material is affected by the presence of air void [11,21]. Research by Zhu et al. showed that the presence of air void has limited influence on the shear wave propagation, and the shear wave velocities are almost consistent at the initial setting [5,21,22], and this consistency was verified on cement pastes and mortar samples. Liu et al. [5] used an embedded bender element to monitor the shear wave velocity change during the setting of mortar and concrete samples, and a strong correlation between the shear wave velocity and penetration resistance was found. Carette determined the setting time of mortars containing two types of fly ashes based on the combined monitoring of P and shear waves, and it was found that shear wave velocity and dynamic elastic properties are the most accurate indicators of the setting process [23]. However, when it comes to the complexity of the implementation of the measurement system, the techniques above seem not that attractive.

Ultrasonic guided waves are elastic waves propagating in a plate or rod. Due to its long propagation range, the guided wave is frequently used in material characterization and damage detection in plates and pipes [24–28]. For instance, Zima and Kedra [27] carried out a series of numerical studies to find out the effect of a concrete mesostructure on lamb wave propagation in concrete plates, and it was found that the displacements associated with wave motion are affected by the mesostructure of the concrete plates. Lee et al. [28] formed an embedded guided wave sensor system by placing two pairs of collocated identical piezoelectric patches on a steel plate with uniform thickness. The system was then used for the hardening process monitoring of the ultra-high performance concrete. It was found that two features of the guided wave, namely the amplitude attenuation and the time-of-flight, are more suitable to monitor the behavior of the ultra-high performance concrete. When a rod is embedded in another medium, part of the guided wave energy will propagate through the interface, and leaky waves are excited in the surrounding medium. The attenuation caused by the leakage depends on the properties of the surrounding materials [29]. Thus, it is possible to evaluate the properties of the surrounding materials by monitoring the attenuation change. The through-transmission method was adopted by Sharma and Mukherjee [30,31] to measure guided waves in a rebar, and the signal amplitude was used to correlate with P wave velocity and compressive strength in concrete. Ervin et al. [32] used this method to monitor corrosion of rebar embedded in mortar. However, this method is not applicable when only one side could be accessed. Vogt et al. [29] investigated the scattering of ultrasonic guided waves at a point where a free cylindrical waveguide enters an embedding material, and the lowest-order longitudinal mode was recommended for embedded guided wave monitoring. As an example, Vogt et al. [33] used a steel wire for guide wave propagation, in a reflected manner, to monitor the curing process of epoxy resins, and the reflected guided wave signals were analyzed using both the reflection coefficient and attenuation method. It was found that both methods are sensitive to the shear properties at low frequencies. Sun and Zhu [34] improved the dispersion calculation of the embedded rebar and used the embedded rebar for guide wave propagation to monitor the early age properties of cement and mortar samples, and in this study the guided wave was generated using an electric coil and received by a commercial ultrasonic transducer. However, to the authors' knowledge, only a steel rebar or wire with a fixed diameter was investigated in the studies above; it would be interesting to make comparisons between rods of different materials and diameters.

In this paper, a series of experimental tests on mortar and concrete samples with various mix designs were performed. Four different kinds of metallic rod were tested simultaneously to find out the optimal guided wave setup to monitor the setting process. Meanwhile, shear wave velocities of the mortar and concrete were monitored, and the time of setting was also measured from these samples. The relationship between these measurements was also discussed.

2. Materials and Methods

2.1. Materials

Two types of samples, namely mortar and concrete samples, were mixed, and ordinary Portland cement was used in both types of samples. The fine aggregate used in the two types of samples was river sand, and the relative density of this sand was 2.62. The coarse aggregate used in concrete samples was a dolomitic limestone with a relative density of

2.65. As shown in Table 1, the mortar samples were designed to span three water to cement ratios (w/c = 0.4, 0.45 and 0.5), and the w/c of the concrete sample was fixed to 0.5 to figure out the influence of coarse aggregate on the guided wave technique.

Table 1. Mixture designs and setting times of mortar and concrete samples.

w/c	Sample Type	Coarse Aggregate Volume (%)	Fine Aggregate Volume (%)	Initial Setting Time (min)	Final Setting Time (min)
0.40	mortar	–	51.1	232	342
0.45	mortar	–	51.1	264	371
0.50	mortar	–	51.1	297	399
0.50	concrete	43.4	28.9	281	384

–not applicable.

2.2. Ultrasonic Test Setup

To find out the optimal material with appropriate mechanical properties and dimensions, four different metallic rods (namely steel rods with a diameter of 3.17 mm and 6.35 mm, and aluminum rods with a diameter of 3.17 mm and 6.35 mm) were embedded into the samples during the tests (see Figure 1a). The length of each rod was 304.8 mm, and the embedded length was 152.4 mm. A Plexiglas plate with four holes was used to keep the rods in place and to reduce the moisture evaporation as well. Low cost small piezo discs (SMD07T02S412, manufactured by STEINER & MARTINS, Inc., Miami, FL. USA) (see Figure 1a,c) bonded on top of the rods were used to excite the rods in the axial direction and to receive the reflected signals. In order to get the signals from the four rods alternately, an Agilent multiplexer 34970A with the plug-in 34903A card (Agilent, Santa Clara, CA, USA) (see Figure 1b) was used to connect these four piezo discs to a pulser-receiver (Olympus 5077PR, Tokyo, Japan) and a digitizer (NI-PXI 5133, Austin, TX, USA). The 34903A card is simply a set of 20 independent single-pole, double-throw reed relays, and each of the 20 relays could be independently controlled. In this study, only four relays were used and closed alternately to get the signals from the four rods. In order to reduce the noise ratio, a total of 200 signals obtained from each rod were averaged separately and then saved each time.

Meanwhile, shear wave testing was carried out using two shear wave transducers (Olympus V151, Olympus, Tokyo, Japan) in the same batch of samples independently. The setup for shear wave testing was the same as the one used in the author's previous work [5], as shown in Figure 1d, the mold was formed by two plastic plates on the sides and a rubber container in between. A soft foam layer was installed between the internal container and external plastic plate, aiming to reduce the direct transmitted ultrasonic waves. Two shear wave transducers were put on each side of the mold. In this setup, the sample thickness was about 27 mm, and the actual thicknesses will be tested after shear wave testing. During the shear wave testing, one transducer served as an actuator, and the other was used as a receiver. Both transducers were connected to an Olympus 5077PR pulser-receiver (Olympus, Tokyo, Japan), with a gain of 40 dB. A 200 V square wave pulse sent from the pulser-receiver was used to drive the actuating transducer. The receiving signals were then sampled at a 10 MHz sampling rate by a NI-USB 5133 digitizer (National Instruments, Austin, TX, USA), which was also used to transfer digitized signals to the connected computer. Furthermore, 200 signals were first obtained and then averaged, as used in the guided wave testing setup, to reduce the noise ratio. Both shear wave velocity tests and guided wave tests were carried out with a 5 min interval until the final setting of the samples.

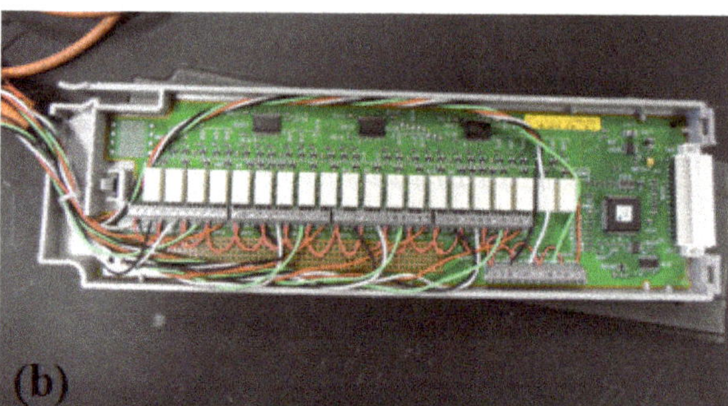
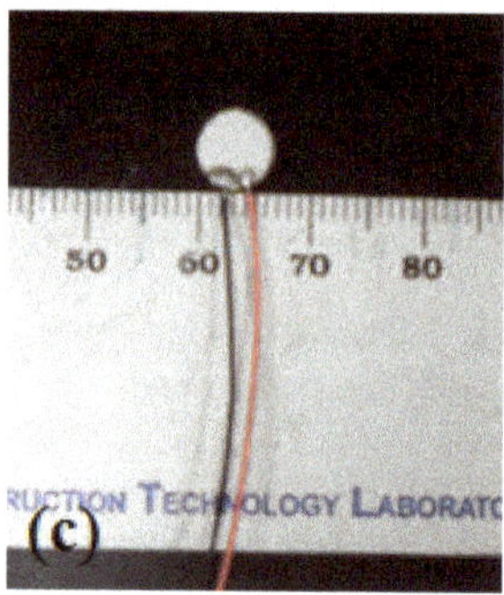
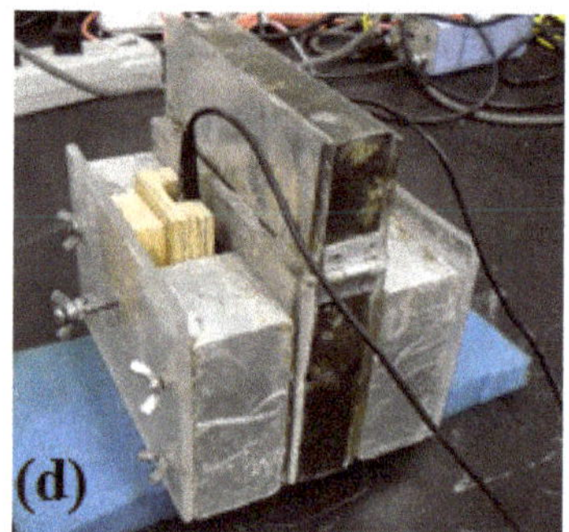

Figure 1. Ultrasonic Test setups: (**a**) guided wave testing; (**b**) 34903A switching card; (**c**) 7 mm diameter piezo disc and (**d**) shear wave measurement.

2.3. Penetrometer Tests

In order to determine the initial and final setting of the samples, penetration resistance tests were carried out on the same batch of mortar and concrete samples, according to ASTM C403 [4]. For concrete samples, sieving is needed, and the sieved mortar was obtained by wet-sieving the selected portion of concrete through a 4.75-mm sieve and onto a non-absorptive surface. In the penetration resistance tests, mortar samples were penetrated using standard needles, and the resistance value was obtained at regular time intervals. Then a plot of penetration resistance versus time was used to determine the times of initial and final setting. The initial and final times of setting correspond to penetration resistance values of 3.5 MPa and 27.6 MPa, respectively. The setting times of all the mixtures are also listed in Table 1.

3. Results and Discussions

3.1. Shear Wave Velocity Testing

It has been shown that the propagation of the shear wave is less affected by the presence of air void in cementitious materials. Thus, only the shear wave velocity is obtained and analyzed in this study. It was noticed that it is hard to pick the first arrival of shear waves from a single piece of signal in the time domain. Thus, B scan images were formed by a stack of signals in the time domain together and used to capture the shear wave arrival. The shear wave arrival time was first obtained using a MATLAB (R2019b, The MathWorks, Inc., Natick, MA, USA) input function to manually pick arrival points at different ages along the arrival time trend. Then, the shear wave velocity was calculated by dividing the thickness of the mortar sample, which was 27 mm in this study, over the travel time of the wave through the sample. Note that a test system time of 4 µs was subtracted to obtain the actual travel time.

Figure 2 shows a B scan image of wave propagation evolvement during setting in a mortar sample (w/c = 0.40), the x-axis and y-axis in the figure representing the time of signals and the age of mortar, respectively. It was noticed that at the first 2.5 h of setting, direct transmission waves transmit along the rubber mold are stronger than the waves propagating directly through the sample. However, by using B scan images, the first arrivals at a very early age can still be obtained.

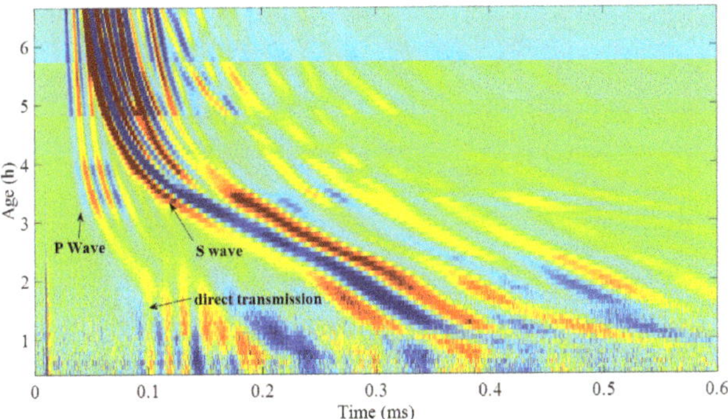

Figure 2. B scan image formed from signals getting from a mortar sample (w/c = 0.40).

Shear wave velocities obtained from different w/c mortar samples are shown in Figure 3, and the times of the initial setting are also marked on the corresponding curves. It is clearly seen that as w/c increases, a much longer initial setting time is taken, and it was also noticed that even though the w/c are different, the shear wave velocities are very consistent (around 410 m/s) at initial setting times. This trend agrees very well with previous research on cement paste [22].

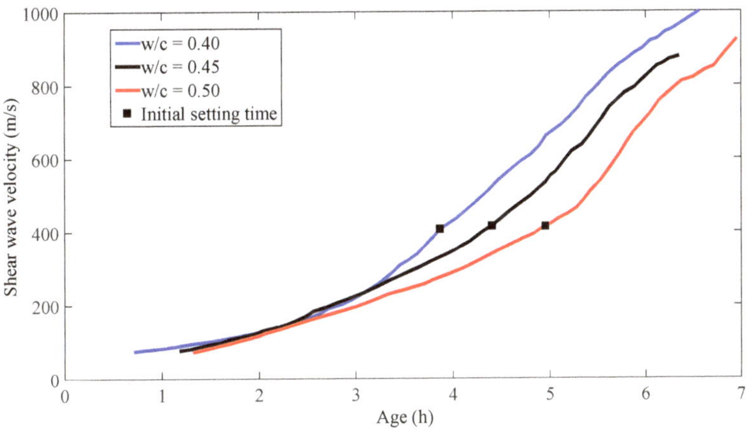

Figure 3. Shear wave velocities in different w/c mortar samples.

3.2. Guided Wave Testing

Many guided wave modes exist in a cylindrical rod, and each mode travels at a different velocity. These wave modes are often categorized into three different types: longitudinal (L), torsional (T) and flexural (F). Figure 4 shows the dispersion curves of the phase velocity for the first few modes in the 6.35 mm aluminum rod. The fundamental L (0, 1) mode was identified as most suitable for this study for several reasons. First of all, at the testing point (see Figure 4), there is mainly axial surface displacement and little radial displacement in the rod, which makes the attenuation very sensitive to the shear wave velocity of the surrounding mortar or concrete. Furthermore, most importantly, the L (0, 1) mode could be easily excited by the piezo discs adopted in this study and is non-dispersive at low frequency. The attenuation caused by the surrounding mortar or concrete could be calculated from:

$$\alpha = -\frac{1}{D} \ln f_0 \left(\frac{A_{emb}}{A_0} \right) \tag{1}$$

where D is the embedded depth, A_0 is the signal amplitude before the rod was embedded, and A_{emb} is the signal amplitude after the rod was embedded.

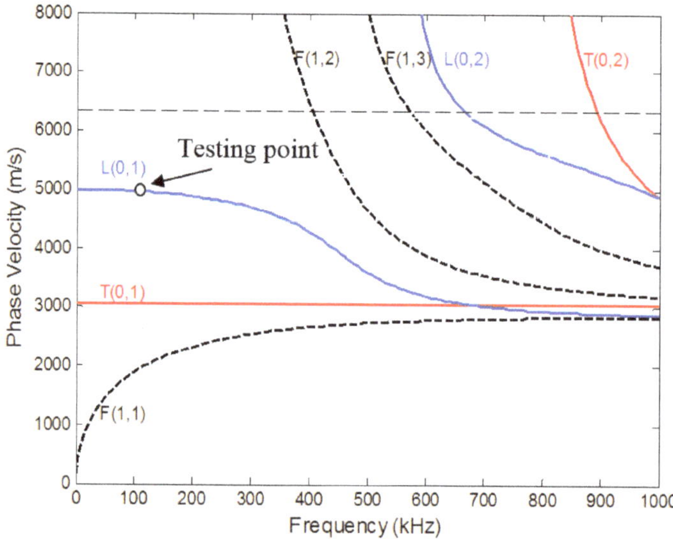

Figure 4. Phase velocity dispersion curves for an aluminum rod of 6.35 mm diameter.

Figure 5 shows two sets of representative waveforms for a mortar sample with w/c of 0.40 using different rods, and only the first two reflection echoes were shown. As shown in Figure 5, amplitudes of the signals getting from the steel rod of 3.17 mm diameter decrease very slowly in the first 4 h. In contrast, the amplitudes of the signals getting from the aluminum rod of 3.17 mm diameter decrease much faster in the first 4 h. This is probably due to the reason that acoustic impedance mismatch between early age mortar and aluminum is much smaller than the mismatch between early age mortar and steel. Thus, more energy is leaked to the surrounding mortar sample in the aluminum rod. This also implies that the aluminum rod is probably more sensitive and suitable to monitor the evolution of early age properties of cementitious samples.

The amplitude of the reflected guided waves was then normalized to the amplitudes of the waves obtained when the rods were not embedded. Figure 6a shows the normalized peak amplitudes getting from four different rods as a function of mortar age. It was noticed that during the first 3 h, the peak amplitudes of both the 6.35 and 3.17 mm steel rods decrease very slowly. As a result, the guided wave attenuation in these two rods does not increase significantly in the first 3 h (see Figure 6b).

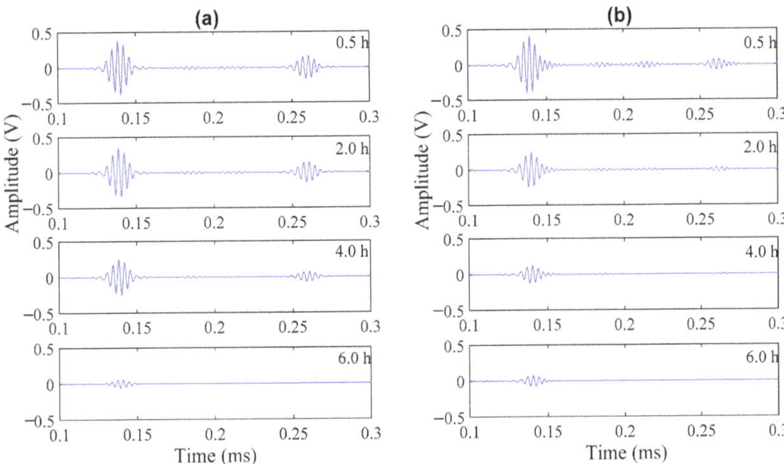

Figure 5. Waveforms at 0.5, 2.0, 4.0, and 6.0 h for a mortar sample with w/c = 0.40 using: (**a**) a steel rod of 3.17 mm diameter; (**b**) an aluminum rod of 3.17 mm diameter.

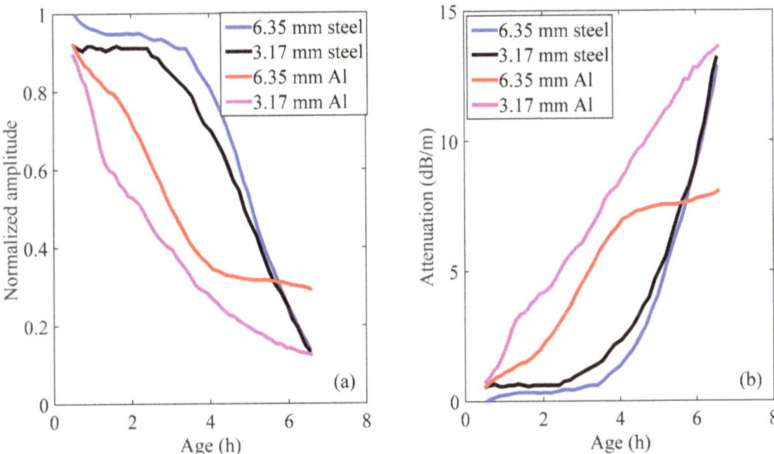

Figure 6. Four different rods in mortar sample with w/c = 0.40. (**a**) normalized wave amplitude and (**b**) attenuation.

Another phenomenon noticed is that the attenuation trends in both steel rods are almost the same, especially after 5 h of setting. The reason was further investigated, and wavelet analysis was adopted to see the time-frequency domain difference between the signals obtained from the 6.35 mm and 3.17 mm diameter steel rods, respectively. As seen in Figure 7a, the central frequency of the reflected echoes obtained from the 6.35 mm diameter steel rod is 135 kHz, corresponding to a frequency-radius product of 428.63 kHz-mm. Sun [34] has proposed an improved approach to calculate the group velocity dispersion curve of rebar embedded into concrete, based on the PCDISP package [35]. By using Sun's approach, the theoretical group velocity in steel rod with a frequency-radius product of 428.63 kHz-mm was found to be 4989 m/s. While the group velocity calculated using the experimental data in this study was 2 × 304.8 mm/(0.2623–0.1403) ms = 4997 m/s, this agrees well with the theoretical group velocity.

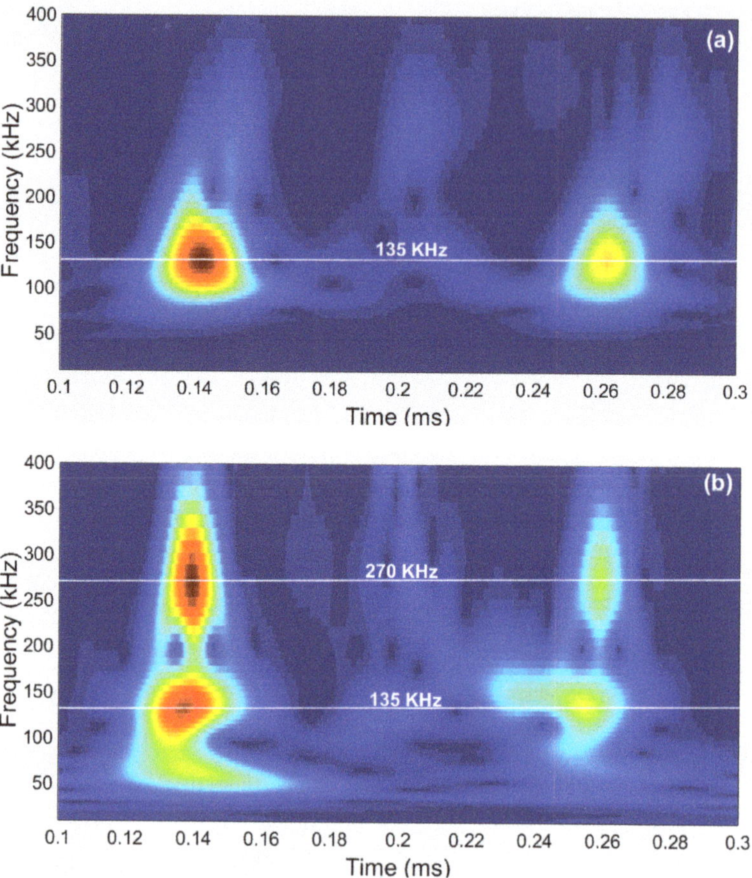

Figure 7. The wavelet of signals obtained from mortar sample with w/c = 0.40 at the age of 0.5 h. (**a**) steel rod with a diameter of 6.35 mm and (**b**) steel rod with a diameter of 3.17 mm.

In the reflected echoes obtained from the 3.17 mm diameter steel rod, two dominant central frequencies, 135 kHz and 270 kHz could be observed. Note that the frequency-radius product of the 270 kHz signal is 428.63 kHz-mm in the 3.17 mm rod, and this product is the same as the one obtained from the 6.35 mm rod. As the hardening process develops, the 428.63 kHz-mm product will gradually become the dominant one in the 3.17 mm rod. Thus, the attenuation trends in both the 6.35 mm and 3.17 mm diameter steel rods are almost the same after 5 h of setting.

As seen in Figure 6, both the 6.35 mm and 3.17 mm aluminum rods show a rapid decrease in peak amplitudes in the first 4 h. Note that from around four hours, the decrease rate of the peak amplitude of the 6.35 mm aluminum rod becomes very slow. While the decrease rate of the peak amplitude of the 3.17 mm aluminum rod remains pretty high, this continuous rapid change makes the 3.17 mm aluminum rod the best choice within the four rods for monitoring the setting of cementitious materials. For simplicity, only the results obtained from the 3.17 mm aluminum rod will be shown in the following sections.

Since there is mainly axial surface displacement and little radial displacement under the low- frequency L (0,1) mode [36], shear leakage will control the attenuation caused by the embedded materials. Therefore, this mode is particularly sensitive to the shear properties of the embedded materials. Note that the shear wave velocity was obtained during the test as well, it would be very interesting to correlate it with the guided wave attenuation.

Figure 8 shows the correlation formed by shear wave velocity versus the guided wave attenuation. As seen in the figure, despite some variations in the very early age, shear wave velocity still correlates strongly with the guided wave attenuation. Specially, after the very early age, there is an approximately linear relationship between these two measurements. Similar relationships, obtained by using steel rod for guide wave propagation, were also found in the hardening process monitoring of epoxy resins [33] and cement pastes [34].

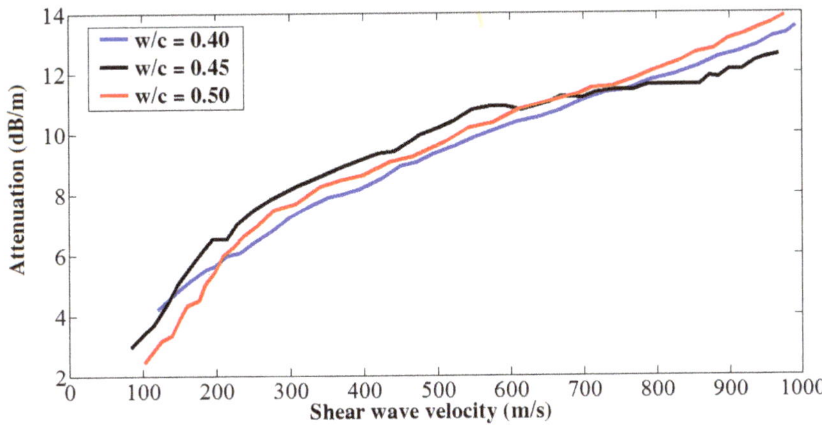

Figure 8. Correlation between shear wave velocity and attenuation.

To find out the influence of coarse aggregates on this technique, a w/c = 0.50 concrete sample was tested in this study. The normalized amplitude and attenuation change during hydration of a concrete sample, and the mortar sample sieved from concrete are shown in Figure 9. As shown in the figure, the developing trends in both materials are almost identical. This is probably because the rod is very thin compared to the size of coarse aggregates, and the rod is still mainly surrounded by mortar when embedded in the concrete sample. This implies that this technique is able to eliminate the effects of coarse aggregates, which makes it of great potential to be applied to the fresh concrete in the field after a standard setup is established. Another thing that needs to be noted is that an increasing practice in concrete engineering is the use of additives to reduce the water dosage, and the rheological properties of concrete samples using additives may be quite different. Thus, extensive attention should be paid when applying this approach to concrete samples using additives.

In order to find out whether coarse aggregates affect the development and the evolution of shear waves in concrete, shear wave velocity change during the setting process is also shown here (see Figure 10). As seen in the figure, the shear wave velocity in concrete is around 860 m/s at the initial setting time, and it is about 450 m/s higher than the shear wave velocity in mortar sieved from the same batch of concrete. This is understandable, because when taking shear wave velocity measurement on concrete samples, the shear wave has to propagate through the coarse aggregates, whose mechanical properties are very different from the surrounding mortar. In other words, the effects of coarse aggregates could not be ignored when taking shear wave velocity measurements.

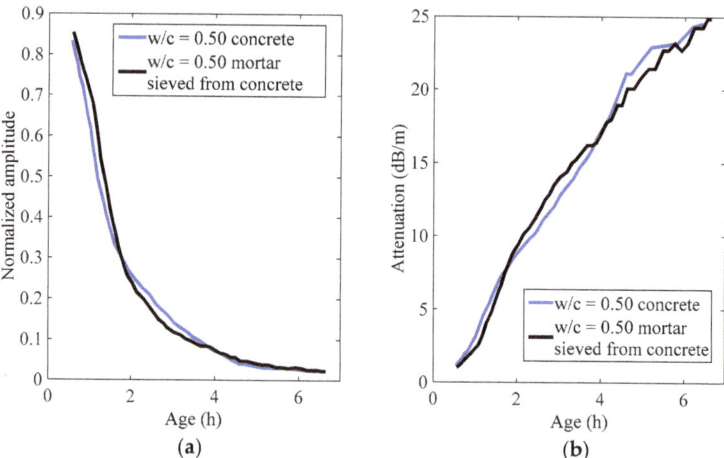

Figure 9. Guided waves in concrete and mortar sieved from concrete: (**a**) normalized amplitude and (**b**) attenuation.

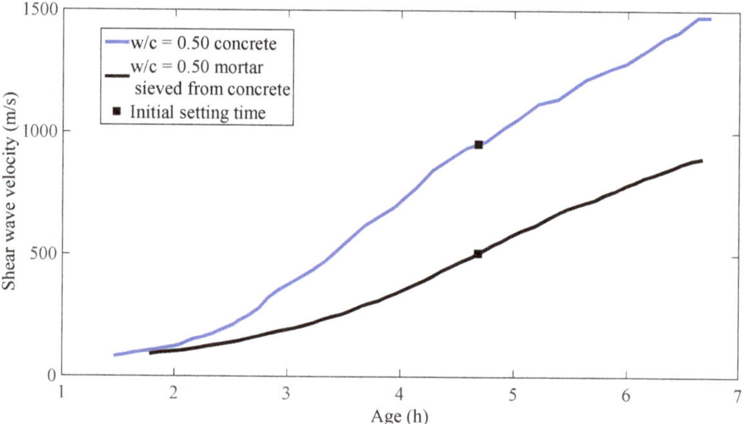

Figure 10. Shear wave velocities in concrete and mortar sieved from concrete.

4. Conclusions

In this study, an embedded guided wave technique has been adopted to monitor the setting process of cementitious samples. Specially, four different kinds of rods have been tested simultaneously to find out the optimal guided wave setup. Shear wave velocities of the mortar and concrete have been monitored and the time of setting has also been measured on these samples, and the relationship between these measurements has been discussed. The following conclusions could be drawn from this study.

Within the four rods tested in this study, the aluminum rod with a diameter of 3.17 mm performs the best. It could be adopted to characterize the whole setting process of cementitious samples with standard water to cement ratios.

By using an aluminum rod in the fundamental mode, a strong correlation between the attenuation and the shear wave velocity has been shown in mortar samples. This trend agrees well with the previous embedded guide wave monitoring studies, which used a steel rod to guide wave propagation.

This technique worked equally well in concrete samples and mortar samples sieved from the same batch of concrete. This implies that this technique is able to eliminate the effects of coarse aggregates, and is of great potential for field setting monitoring of cementitious materials with standard water to cement ratios.

Author Contributions: Methodology, G.Y. and S.L.; analysis, D.W. and S.L.; validation, D.W. and P.S.; writing—original draft preparation, S.L.; writing—review and editing, G.Y., D.W. and P.S.; funding acquisition, G.Y. and S.L. All authors have read and agreed to the published version of the manuscript.

Funding: National Key Research and Development Project (2019YFE0118500) and Academic Frontier Project of China University of Mining and Technology (2017XKQY050) are greatly appreciated for their generous funding.

Institutional Review Board Statement: Not applicable.

Informed Consent Statement: Not applicable.

Data Availability Statement: Data available on request due to their size properties. The data presented in this study are available on request from the corresponding author.

Conflicts of Interest: The authors declare no conflict of interest.

References

1. Javier Vazquez-Rodriguez, F.; Elizondo-Villareal, N.; Hypatia Verastegui, L.; Arato Tovar, A.M.; Fernando Lopez-Perales, J.; de Leon, J.E.; Gomez-Rodriguez, C.; Fernandez-Gonzalez, D.; Felipe Verdeja, L.; Viviana Garcia-Quinonez, L.; et al. Effect of Mineral Aggregates and Chemical Admixtures as Internal Curing Agents on the Mechanical Properties and Durability of High-Performance Concrete. *Materials* **2020**, *13*, 2090. [CrossRef] [PubMed]
2. Kim, S.S.; Qudoos, A.; Jakhrani, S.H.; Lee, J.B.; Kim, H.G. Influence of Coarse Aggregates and Silica Fume on the Mechanical Properties, Durability, and Microstructure of Concrete. *Materials* **2019**, *12*, 3324. [CrossRef] [PubMed]
3. Long, W.-J.; Khayat, K.H.; Lemieux, G.; Hwang, S.-D.; Han, N.-X. Performance-Based Specifications of Workability Characteristics of Prestressed, Precast Self-Consolidating Concrete-A North American Prospective. *Materials* **2014**, *7*, 2474–2489. [CrossRef] [PubMed]
4. ASTM C403/C403M-08. *Standard Test Method for Time of Setting of Concrete Mixtures by Penetration Resistance*; ASTM International, 100 Barr Harbor Drive: West Conshohocken, PA, USA, 2008. [CrossRef]
5. Liu, S.; Zhu, J.; Seraj, S.; Cano, R.; Juenger, M. Monitoring setting and hardening process of mortar and concrete using ultrasonic shear waves. *Constr. Build. Mater.* **2014**, *72*, 248–255. [CrossRef]
6. Boumiz, A.; Vernet, C.; Tenoudji, F.C. Mechanical properties of cement pastes and mortars at early ages: Evolution with time and degree of hydration. *Adv. Cem. Based Mater.* **1996**, *3*, 94–106. [CrossRef]
7. Chotard, T.; Gimet-Breart, N.; Smith, A.; Fargeot, D.; Bonnet, J.P.; Gault, C. Application of ultrasonic testing to describe the hydration of calcium aluminate cement at the early age. *Cem. Concr. Res.* **2001**, *31*, 405–412. [CrossRef]
8. D'Angelo, R.; Plona, T.J.; Schwartz, L.M.; Coveney, P. Ultrasonic measurements on hydrating cement slurries: Onset of shear wave propagation. *Adv. Cem. Based Mater.* **1995**, *2*, 8–14. [CrossRef]
9. Keating, J.; Hannant, D.J.; Hibbert, A.P. Comparison of shear modulus and pulse velocity techniques to measure the build-up of structure in fresh cement pastes used in oil well cementing. *Cem. Concr. Res.* **1989**, *19*, 554–566. [CrossRef]
10. Sant, G.; Ferraris, C.F.; Weiss, J. Rheological properties of cement pastes: A discussion of structure formation and mechanical property development. *Cem. Concr. Res.* **2008**, *38*, 1286–1296. [CrossRef]
11. Sayers, C.M.; Dahlin, A. Propagation of ultrasound through hydrating cement pastes at early times. *Adv. Cem. Based Mater.* **1993**, *1*, 12–21. [CrossRef]
12. Ye, G.; van Breugel, K.; Fraaij, A.L.A. Experimental study and numerical simulation on the formation of microstructure in cementitious materials at early age. *Cem. Concr. Res.* **2003**, *33*, 233–239. [CrossRef]
13. Reinhardt, H.W.; Grosse, C.U. Continuous monitoring of setting and hardening of mortar and concrete. *Constr. Build. Mater.* **2004**, *18*, 145–154. [CrossRef]
14. Sayers, C.M.; Grenfell, R.L. Ultrasonic propagation through hydrating cements. *Ultrasonics* **1993**, *31*, 147–153. [CrossRef]
15. Lee, H.K.; Lee, K.M.; Kim, Y.H.; Yim, H.; Bae, D.B. Ultrasonic in-situ monitoring of setting process of high-performance concrete. *Cem. Concr. Res.* **2004**, *34*, 631–640. [CrossRef]
16. Qin, L.; Li, Z. Monitoring of cement hydration using embedded piezoelectric transducers. *Smart Mater. Struct.* **2008**, *17*, 55005. [CrossRef]
17. Rapoport, J.R.; Popovics, J.S.; Kolluru, S.V.; Shah, S.P. Using ultrasound to monitor stiffening process of concrete with admixtures. *ACI Mater. J.* **2000**, *97*, 675–683.
18. Voigt, T.; Grosse, C.U.; Sun, Z.; Shah, S.P.; Reinhardt, H.W. Comparison of ultrasonic wave transmission and reflection measurements with P-and S-waves on early age mortar and concrete. *Mater. Struct.* **2005**, *38*, 729–738. [CrossRef]

19. Dumoulin, C.; Karaiskos, G.; Carette, J.; Staquet, S.; Deraemaeker, A. Monitoring of the ultrasonic P-wave velocity in early-age concrete with embedded piezoelectric transducers. *SMART Mater. Struct.* **2012**, *21*, 047001. [CrossRef]
20. Zhu, J.; Cao, J.N.; Bate, B.; Khayat, K.H. Cement and Concrete Research Determination of mortar setting times using shear wave velocity evolution curves measured by the bender element technique. *Cem. Concr. Res.* **2018**, *106*, 1–11. [CrossRef]
21. Zhu, J.; Kee, S.-H.; Han, D.; Tsai, Y.-T. Effects of air voids on ultrasonic wave propagation in early age cement pastes. *Cem. Concr. Res.* **2011**, *41*, 872–881. [CrossRef]
22. Zhu, J.; Tsai, Y.-T.; Kee, S.-H. Monitoring early age property of cement and concrete using piezoceramic bender elements. *Smart Mater. Struct.* **2011**, *20*, 115014. [CrossRef]
23. Carette, J.; Staquet, S. Monitoring the setting process of mortars by ultrasonic P and S-wave transmission velocity measurement. *Constr. Build. Mater.* **2015**, *94*, 196–208. [CrossRef]
24. Li, F.; Murayama, H.; Kageyama, K.; Shirai, T. Guided Wave and Damage Detection in Composite Laminates Using Different Fiber Optic Sensors. *Sensors* **2009**, *9*, 4005–4021. [CrossRef] [PubMed]
25. Ghavamian, A.; Mustapha, F.; Baharudin, B.T.H.T.; Yidris, N. Detection, Localisation and Assessment of Defects in Pipes Using Guided Wave Techniques: A Review. *Sensors* **2018**, *18*, 4470. [CrossRef]
26. Radzienski, M.; Kudela, P.; Marzani, A.; De Marchi, L.; Ostachowicz, W. Damage Identification in Various Types of Composite Plates Using Guided Waves Excited by a Piezoelectric Transducer and Measured by a Laser Vibrometer. *Sensors* **2019**, *19*, 1958. [CrossRef]
27. Zima, B.; Kedra, R. Numerical Study of Concrete Mesostructure Effect on Lamb Wave Propagation. *Materials* **2020**, *13*, 2570. [CrossRef]
28. Lee, C.; Park, S.; Bolander, J.E.; Pyo, S. Monitoring the hardening process of ultra high performance concrete using decomposed modes of guided waves. *Constr. Build. Mater.* **2018**, *163*, 267–276. [CrossRef]
29. Vogt, T.; Lowe, M.; Cawley, P.; Introduction, I. The scattering of guided waves in partly embedded cylindrical structures. *J. Acoust. Soc. Am.* **2013**, *113*, 1258–1272. [CrossRef]
30. Sharma, S.; Mukherjee, A. Cement & Concrete Composites Monitoring freshly poured concrete using ultrasonic waves guided through reinforcing bars. *Cem. Concr. Compos.* **2015**, *55*, 337–347. [CrossRef]
31. Sharma, S.; Mukherjee, A. Ultrasonic guided waves for monitoring the setting process of concretes with varying workabilities. *Constr. Build. Mater.* **2014**, *72*, 358–366. [CrossRef]
32. Ervin, B.L.; Kuchma, D.A.; Bernhard, J.T.; Reis, H. Monitoring Corrosion of Rebar Embedded in Mortar Using High-Frequency Guided Ultrasonic Waves. *J. Eng. Mech.* **2009**, *135*, 9–19. [CrossRef]
33. Vogt, T.; Lowe, M.; Cawley, P. Cure monitoring using ultrasonic guided waves in wires. *J. Acoust. Soc. Am.* **2004**, *114*, 1303–1313. [CrossRef] [PubMed]
34. Sun, H.; Zhu, J. Monitoring Early Age Properties of Cementitious Material Using Ultrasonic Guided Waves in Embedded Rebar. *J. Nondestruct. Eval.* **2017**, *36*, 1–12. [CrossRef]
35. Seco, F.; Jiménez, A.R.; Automática, C. De Modelling the Generation and Propagation of Ultrasonic Signals in Cylindrical Waveguides. In *Ultrasonic Waves*; dos Santos, A.A., Jr., Ed.; IntechOpen: London, UK; Available online: https://www.intechopen.com/books/ultrasonic-waves/modelling-the-generation-and-propagation-of-ultrasonic-signals-in-cylindrical-waveguides (accessed on 18 January 2021). [CrossRef]
36. Pu, S.H.; Cegla, F.; Drozdz, M.; Lowe, M.J.S.; Cawley, P.; Buenfeld, N.R. Monitoring the setting and early hardening of concrete using an ultrasonic waveguide. *Insight* **2004**, *46*, 350–354. [CrossRef]

materials

Article

Disassembly Study of Ultrasonically Welded Thermoplastic Composite Joints via Resistance Heating

Harry Frederick, Wencai Li and Genevieve Palardy *

Department of Mechanical and Industrial Engineering, Louisiana State University, 3261 Patrick F. Taylor Hall, Baton Rouge, LA 70803, USA; hfrede6@lsu.edu (H.F.); wli45@lsu.edu (W.L.)
* Correspondence: gpalardy@lsu.edu

Abstract: This manuscript explores the disassembly potential of ultrasonically welded thermoplastic composite joints for reuse or recycling through resistance heating via a nanocomposite film located at the welded interface. Nanocomposite films containing multi-walled carbon nanotubes (MWCNTs) were characterized for thermo-electrical behavior to assess self-heating. It was generally observed that maximum temperature increased with MWCNT and film thickness. To demonstrate potential for disassembly, glass fiber/polypropylene adherends were welded with nanocomposite films. Shear stress during disassembly was measured for three initial adherend's surface temperatures. It was found that the required tensile load decreased by over 90% at the highest temperatures, effectively demonstrating the potential for disassembly via electrically conductive films. Fracture surfaces suggested that disassembly was facilitated through a combination of nanocomposite and matrix melting and weakened fiber–matrix interface. Limitations, such as slow heating rates and the loss of contact at the interface, imply that the method could be more suited for recycling, instead of repair and reuse, as the heat-affected zone extended through the adherends' thickness at the overlap during heating.

Keywords: thermoplastic composites; ultrasonic joints; resistance heating

check for **updates**

Citation: Frederick, H.; Li, W.; Palardy, G. Disassembly Study of Ultrasonically Welded Thermoplastic Composite Joints via Resistance Heating. *Materials* **2021**, *14*, 2521. https://doi.org/10.3390/ma14102521

Academic Editor: Francesca Lionetto

Received: 19 April 2021
Accepted: 10 May 2021
Published: 12 May 2021

Publisher's Note: MDPI stays neutral with regard to jurisdictional claims in published maps and institutional affiliations.

1. Introduction

Thermoplastic composites (TPCs) are used in several industries, such as automotive, aerospace, and wind energy, because of their high specific modulus and strength, fracture toughness, damage tolerance, and impact and corrosion resistance [1–4]. Common thermoplastic matrices include polypropylene (PP), polyethylene (PE), and polycarbonate (PC), used in a range of low-cost applications. High-performance thermoplastic matrices encompass higher temperature polymers, such as polyamide 6 (PA6), polyetherimide (PEI), polyphenylene sulfide (PPS), and polyether ether ketone (PEEK) [5]. In addition, Arkema recently developed a liquid thermoplastic resin with similar mechanical properties to thermosetting resins, Elium®, which enables the use of composite manufacturing technologies traditionally used for thermosets [6]. As thermoplastics can be thermoformed when heated up above certain temperatures, TPCs have the potential for recycling, reuse and reshaping into new components, as well as joining through fusion bonding [7]. The latter can eliminate the use of rivets, reducing weight, cost and stress concentration. It is more time-efficient than thermoset adhesive bonding because it does not require surface preparation. Fusion bonding is categorized into thermal, electromagnetic and friction welding [3]. Ultrasonic welding (USW) is a technique that has gained momentum in the past few years for TPCs, as it is fast, energy-efficient, and suitable for spot and continuous joining configurations [8–11].

USW joins adherends by the application of high frequency, low amplitude vibrations through a sonotrode (or horn) to generate heat via frictional and viscoelastic mechanisms [12,13]. An "energy director" (ED) must usually be placed at the weld interface to concentrate heat generation. Triangular protrusions are typically employed in the plastics industry, but for continuous fiber-reinforced thermoplastics, thin films are also suitable and lead to high strength welds [14–18]. Many studies in the literature experimentally investigated the effect of process parameters (amplitude, force and control mode) and ED geometry on bond quality [10,14,15,19–26] and heat generation [18,27,28]. For instance, it was reported that using the vertical displacement of the sonotrode could lead to consistent weld quality using power and displacement curves from the welder. On the other hand, energy director-less welding was found to be possible when controlling the process through time. The prediction of temperature profiles, consumed power, ED flow, and bond strength has shown reasonable accuracy through multi-physics modeling and artificial intelligence methods [13,29–35].

While the USW process and bond strength have been extensively studied for a wide range of TPCs, there is limited research on structural health monitoring and repair of joints. Prior research has shown the potential for multifunctional, nanocomposite films as EDs for USW [36,37]. Those films, rendered electrically conductive by the addition of multi-walled carbon nanotubes (MWCNTs), enabled USW and structural health monitoring at the welded interface through electrical resistance changes. Another function they could fulfill is localized resistance heating at the interface to facilitate disassembly and repair.

Nanocomposite-based heating elements were recently developed and successfully used for the resistance welding of TPCs, notably by Brassard et al. [38,39]. MWCNT/PEI nanocomposite films with weight fractions up to 15 wt.% led to an electrical conductivity of 0.92 S/cm. With 10 wt.% MWCNT, the films reached the glass transition temperature (>217 °C) at an applied voltage of 25 V, demonstrating their Joule heating behavior. However, infrared camera monitoring revealed non-uniform temperature distribution, likely due to copper electrodes acting as heat sinks. In the literature, a wider range of studies on nanocomposite films as susceptors for the induction welding of TPCs or induction heating of adhesives have been carried out. Farahani et al. showed that silver nanoparticle-based thermoplastic films are suitable as susceptors for induction welding, reaching melting temperature in less than 50 s at 400 A [40,41]. However, the potential for disassembly and the repair of fusion bonded TPC joints has not been investigated in the literature.

Although they have not been used to join TPCs, reversible adhesives were developed to facilitate disassembly and the healing of thermoset composite adherends [42,43]. Those adhesives are made of a thermoplastic matrix, containing ferromagnetic nanoparticles to induce temperature increase through induction heating. Reversible joints provide the benefits of both adhesive and mechanically fastened techniques, including ease of disassembly. The method was demonstrated with acrylonitrile butadiene styrene (ABS) and up to 20 wt.% ferromagnetic nanoparticles, but is so far limited to fiber-reinforced epoxy composites. This means the adherends would not be significantly affected during the process. In the fusion bonding of TPCs, the disassembly procedure would be expected to affect the adherends, as the bond line is made of the same thermoplastic as the adherends. To the best of the authors' knowledge, disassembly studies on welded TPC joints have not been reported in the literature. Therefore, the aim of this research work is to address this gap by focusing on two particular topics: (1) assess ease of disassembly for ultrasonically welded joints by investigating effect of resistance heating temperature; (2) understand disassembly mechanisms and the extent to which adherends are affected during the process.

This study will demonstrate the potential for disassembly of ultrasonically welded TPC joints via resistance heating. First, the thermo-electrical characterization of MWCNT-based nanocomposite films containing different filler weight fractions was investigated to assess their use as heating elements. Second, ultrasonic welding was used to assemble glass fiber/polypropylene adherends into a single lap joint configuration. Disassembly was carried out with a tensile testing apparatus under a range of applied voltages at the interface, leading to different interface temperatures. Third, the behavior of the joints during the disassembly procedure was analyzed through shear stress and temperature curves, fractography analysis and extent of heat-affected zone. Finally, the limitations of this technique and proposed future research directions will be discussed.

2. Materials and Experimental Methods

2.1. Materials

Polypropylene (PP) masterbatches containing multi-walled carbon nanotubes (MWCNTs) were used for this study. MWCNT loading ratios equal to 15 wt.%, 20 wt.%, and 25 wt.% were purchased from Cheap Tubes Inc (Brattleboro, VT, USA). Those specific ratios were selected based on preliminary Joule heating experiments, previous work demonstrating suitability for ultrasonic welding [36,37] and literature on high content CNT-based polymer films for sensing, self-heating and resistance welding [38,39,44,45].

Glass fiber/polypropylene (GF/PP) adherends were used for ultrasonic welding and disassembly testing. GF/PP IE 6030 unitape Polystrand™ prepregs with a fiber volume fraction of 60%, an areal weight of 461 g/m^2 and a tape thickness of 0.33 mm were purchased from Avient (formerly PolyOne, Englewood, CO, USA).

2.2. Nanocomposite Films and Thermoplastic Composites Fabrication

The nanocomposite films (MWCNT/PP) were manufactured with a heated laboratory press (Dake, Grand Haven, MI, USA). During compression molding, PTFE (polytetrafluoroethylene)-coated fiberglass release films and steel shims were placed between the heated platens to produce a consistent surface finish and to control the films' final thickness. Based on previous work on the effect of ED thickness on USW of TPCs [10,16,46], shims with thicknesses equal to 0.06 mm, 0.25 mm and 0.50 mm were selected. The molded nanocomposite films were cut into rectangular specimens and stored in sealed bags. For thermo-electrical characterization, the films dimensions were 50 mm × 15 mm, while they were 30 mm × 25 mm for ultrasonic welding.

GF/PP substrates were manufactured by compression molding with a laboratory press (Dake, Grand Haven, MI, USA). Eight unidirectional prepreg layers, measuring 254 mm × 254 mm, were stacked in a $[0]_8$ sequence between steel plates, then placed between the press' heated platens. The laminate was consolidated under 1 MPa at 180 °C for approximately 15 min. During compression molding, a thermocouple was placed at one edge of the laminate to monitor the temperature between the plies. After demolding, a laminate with a final thickness between 1.8 mm and 1.9 mm was obtained. Prior to welding, the laminate was cut into rectangular specimens (101.6 mm × 25.4 mm) with a water-cooled diamond saw (PICO 155 from Pace Technologies). The longer side was cut along the direction of the glass fibers.

2.3. Thermo-Electrical Characterization

Electrical conductivity and the resistance heating measurement setup is illustrated in Figure 1a. A voltage was applied through copper electrodes placed at both ends of the films with a Keithley Sourcemeter 2604B (maximum voltage and current of 40 V and 1 A, respectively), shown in Figure 1b,c. The DC voltages used to gather electrical and thermal data were 1 V, 2 V, 4 V, 6 V, 8 V, and 10 V. All film thicknesses were tested (0.06 mm, 0.25 mm, and 0.50 mm) to assess their effect on thermo-electrical behavior. The electrical conductivity of the films (σ, in S/cm) was calculated using Equation (1), shown below, where R_{Avg} is the average resistance (in Ohms), L is the length of the film between electrodes (in cm), and A is the cross-sectional area of the film (in cm^2):

$$\sigma = \frac{L}{R_{Avg}A} \tag{1}$$

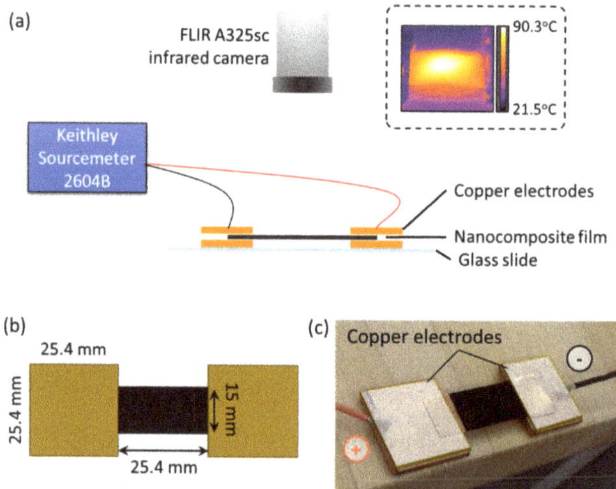

Figure 1. (**a**) Overall setup used for electrical conductivity and Joule heating measurements. Inset shows representative 2D temperature plot recorded with infrared camera at 10 V for a 0.50 mm-thick 20 wt.% MWCNT/PP film after approximately 1 min; (**b**) copper electrodes and nanocomposite film dimensions; (**c**) example of actual nanocomposite film and electrodes placement. Dimensions are not to scale.

For resistance heating, each voltage was applied for three minutes with 30 s between voltages, while the Keithley KickStart software (version 2.0, Beaverton, OR, USA) acquired resistance, power and current data at a rate of 10 data points/second. Two-dimensional temperature plots were acquired with a FLIR A325sc infrared camera (FLIR Systems, Spicewood, TX, USA) placed above the film, at a rate of 15 Hz (example shown in Figure 1a in inset). Temperature profiles were extracted at the center of the film. For each applied voltage, MWCNT weight fraction and film thickness, seven to ten samples were tested.

2.4. Ultrasonic Welding Procedure

GF/PP adherends were welded in a single lap configuration with an overlap area of 25.4 mm × 12.7 mm. A Dynamic 3000 ultrasonic welder (Rinco Ultrasonics, Danbury, CT, USA) with a maximum power of 3000 W and a constant operating frequency of 20 kHz was used with a 40 mm diameter titanium sonotrode. The booster and sonotrode gains were 1:1.5 and 1:3.85, respectively. Both adherends were clamped with aluminum bars and M8 socket head screws on a baseplate, as shown in Figure 2a,b. A nanocomposite film was placed between the adherends to act as the energy director. Films containing 15 wt.%, 20 wt.%, and 25 wt.% MWCNT with a thickness of 0.50 mm were used. Even though thermo-electrical characterization was performed on three different thicknesses, the thickest films were selected for welding and disassembly because they led to the highest bond line thickness, as will be discussed in Section 4.1.1.

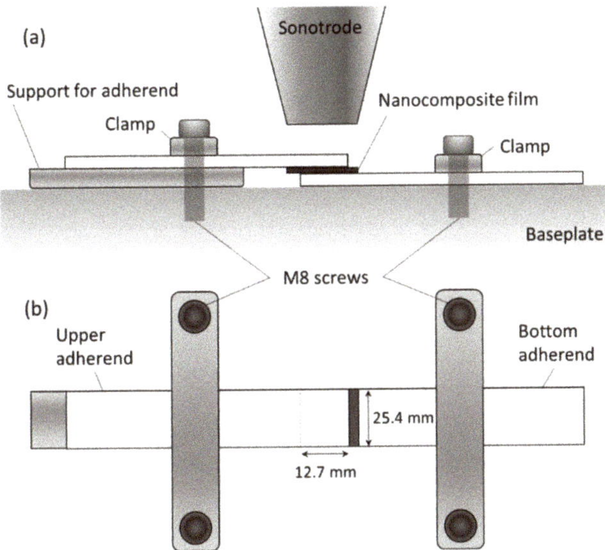

Figure 2. Schematic of ultrasonic welding fixture used in this study for GF/PP adherends and MWCNT/PP energy director films: (**a**) side view and (**b**) top view. Dimensions are not to scale.

For all welds, during the vibration phase, a force of 1000 N and an amplitude of 38.1 µm were applied. The duration of the vibration phase was controlled through the vertical displacement of the sonotrode (also called "travel") until it reached a value equal to 60% of the initial films' thickness. After reaching the prescribed travel value, a consolidation force of 1000 N, for a duration of 4 s, was applied. Those welding parameters were chosen based on previous research [36,37]. The power and travel curves with respect to welding time were acquired for each weld.

2.5. Disassembly of Welded Joints

Following ultrasonic welding, excess polymer at the interface edges was removed with a razor blade. Two 30 AWG copper wires were connected at the interface with silver paint (SPI #05002-AB, electrical resistivity of 1.2×10^{-4} Ohms·cm) to maximize electrical contact. Painted wires and interface were dried overnight before disassembly experiments.

In order to quantify the shear stress required to disassemble the welded joints, the samples were tested with a 50 kN tensile machine (TestResources 313, TestResources Inc., Shakopee, MN, USA), according to ASTM D1002. A schematic of the test setup is shown in Figure 3. The specimens were clamped between hydraulic grips at a distance of 60 mm. The position of both grips was adjusted so that the load direction was aligned with the overlap direction. A voltage between 14 V and 20 V was applied while monitoring the surface temperature of the GF/PP adherend with a FLIR A325sc infrared camera (FLIR Systems, Spicewood, TX, USA). While an external monitoring method does not provide the same accuracy as an embedded sensor at the interface, it was selected for two reasons: (1) the ultrasonic welding process may affect the position and integrity of embedded sensors at the interface due to ultrasonic vibrations (e.g., thermocouples or fiber optic sensors); (2) temperature measurements might become inaccurate as failure initiation and propagation occurs at the interface. Disassembly was initiated at a loading rate of 1.3 mm/min when the surface temperature reached either 110 °C, 130 °C, or 150 °C at the center point of the 25.4 mm × 12.7 mm overlap (delineated by a red, dashed rectangle in Figure 3). Those three temperature values were selected based on simplified 3D thermal analyses, detailed in Section 3 of this manuscript. After disassembly, the extent of the heat-affected zone (HAZ) was quantified using image analysis with the ImageJ software

(National Institutes of Health, NIH, version 1.53e, Bethesda, MD, USA). The HAZ area was measured and its % value was calculated with respect to the overlap area.

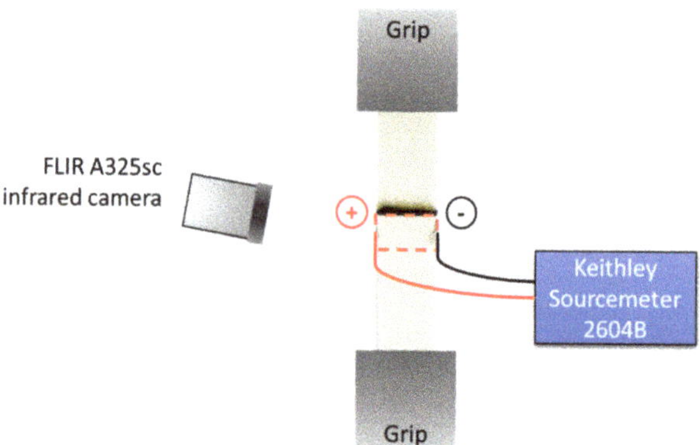

Figure 3. Disassembly setup with infrared camera and sourcemeter. The dashed rectangle indicates the surface area that was monitored for temperature during the tests. Not to scale.

2.6. Scanning Electron Microscopy (SEM)

After disassembly, the microstructure of the fracture surfaces was observed through SEM. Surfaces were coated with gold using a sputter coater (EMS550X, Electron Microscopy Sciences, Hatfield, PA, USA) under a vacuum of 10^{-1} mbar, at 25 mA for 2 min. A high-performance JSM-6610LV SEM (JEOL Ltd., Tokyo, Japan) was employed to capture images, at an acceleration voltage of 15 kV.

3. Prediction of Heat-Affected Zone for Disassembly Experiments

Despite the low thickness of the bond line ($\leq$0.50 mm) and adherends (<1.90 mm), a lag was expected between the surface temperature and the actual temperature at the welded interface. A simplified 3D thermal analysis was carried out in the SolidWorks Thermal Simulation module to predict the temperature at the surface of the adherend (as shown in Figure 3) for various interface temperatures. The goal of these analyses was to provide a range of surface temperatures that would guide the design of the disassembly experiments, based on material properties from the suppliers and found in the literature.

Figure 4a shows the boundary conditions for the thermal finite element analysis (FEA). The thickness of the bond line was 0.1 mm (half-thickness of MWCNT/PP interface) and the GF/PP adherend was 1.8 mm thick. A forced air convection coefficient of 12 W/m^2·K was selected for the room in which tests were to be carried out. It was applied to all adherend surfaces in contact with air. The contact between ED and adherend was defined as "Bonded". The temperature at the interface was set at three values: 120 °C, 140 °C or 160 °C, based on the melting temperature of GF/PP adherends and MWCNT/PP films, between 140 °C and 150 °C [36]. Due to the orthotropic behavior of the adherends (UD layup), two thermal conductivity values (k_y, $k_x = k_z$) were estimated using the rule of mixture shown in Equation (2) and Equation (3):

$$k_y = (1 - V_{GF})k_{PP} + V_{GF}k_{GF}, \qquad (2)$$

$$\frac{1}{k_{x,z}} = \frac{(1 - V_{GF})}{k_{PP}} + \frac{V_{GF}}{k_{GF}} \qquad (3)$$

where V_{GF} is the glass fiber volume fraction, k_{PP} is the thermal conductivity of polypropylene (in W/m·K) and k_{GF} is the thermal conductivity of glass fibers (in W/m·K). The material properties for the GF/PP adherends are listed in Table 1. The heat capacity, C_p (in J/kg·K), was calculated based on the rule of mixture, as described in Equation (2).

The MWCNT/PP films were assumed to exhibit isotropic properties with random carbon nanotubes orientation and distribution. The main thermal properties are listed in Table 1 with the corresponding references in the literature. The nanocomposite films' thermal conductivity was estimated to range between 0.55 W/m·K and 0.65 W/m·K based on [47,48], to account for MWCNT weight fraction and potential variations in dispersion.

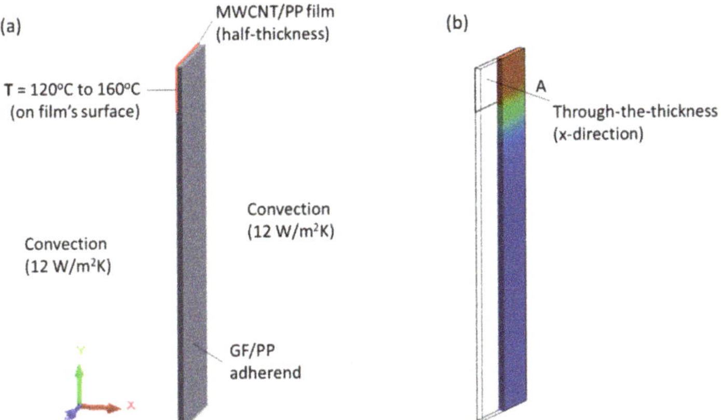

Figure 4. (a) Boundary conditions for finite element analysis used for prediction of temperature profile through the thickness of GF/PP joint. Symmetry was assumed along the ZY plane; and (b) Example of 3D thermal plot with location of plotted results at the mid-plane, along x-direction at center of bond line.

Table 1. Main estimated GF/PP adherend and MWCNT/PP films thermal properties used in FEA. Refer to Figure 4a for coordinate system.

GF/PP					
V_{GF} (%)	k_{GF} (W/m·K)	k_{PP} (W/m·K)	k_y (W/m·K)	k_x, k_z (W/m·K)	C_p (J/kg·K)
60 [a]	1.05 [b]	0.15 [a]	0.69	0.31	1.22
MWCNT/PP					
MWCNT wt (%)	k_{CNT} (W/m·K)	k_{PP} (W/m·K)	$k_{CNT/PP}$ (W/m·K)		C_p (J/kg·K)
15/20/25	3000 [c]	0.15 [a]	0.55 to 0.65 [c]		1.50 [c]

a: Suppliers' specifications sheet (PolyOne and Professional Plastics); b: [5]; c: [47,48].

Figure 5 shows the through-the-thickness temperature profiles along line A at the cross-section labeled in Figure 4b. The results for 15 wt.% MWCNT/PP film at the bond line are presented, but no significant differences were found for the range of $k_{CNT/PP}$ values in Table 1. The surface temperature at the center point of the overlap is 112.8 °C, 131.3 °C and 149.9 °C for an interface temperature of 120 °C, 140 °C and 160 °C, respectively. Given a temperature gradient around 10 °C, as well as the assumptions and simplifications made for thermal analysis, it was estimated that the disassembly experiments should be carried out when the surface temperature of the GF/PP adherend reached 110 °C, 130 °C and 150 °C to adequately capture the behavior of the heated joint.

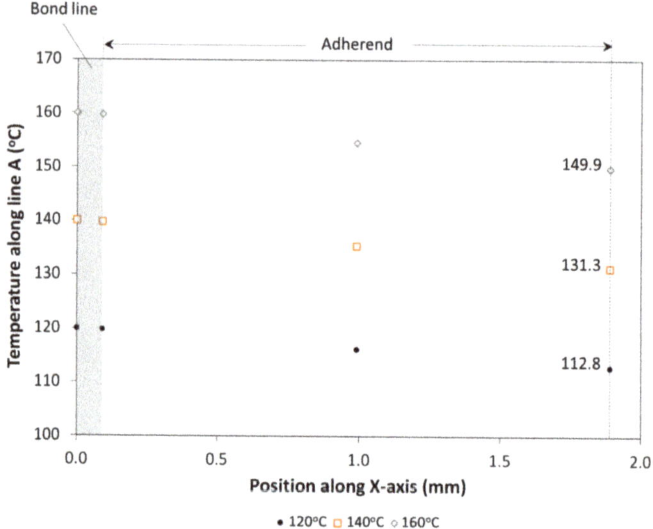

Figure 5. Predicted through-the-thickness temperature profiles along line A at the cross-section labeled in Figure 4b when interface is set at a temperature of 120 °C, 140 °C and 160 °C. The bond line material was 15 wt.% MWCNT/PP.

4. Experimental Results and Discussion

4.1. Nanocomposite Films Characterization

4.1.1. Electrical Conductivity

Figure 6a shows results for electrical conductivity measurements of nanocomposite PP films containing 15 wt.%, 20 wt.%, and 25 wt.% MWCNT, across all three thicknesses. Overall, values are in the same order of magnitude as previously observed in the literature for MWCNT/PP films [49–51], as well as for nanocomposite heating elements designed for resistance welding with weight fractions above 10 wt.% MWCNT [39]. Two general trends are observed. First, for the same applied voltage, average conductivity generally increased with CNT weight fraction, more prominently at 1 V and 2 V. Second, for the same weight fraction, conductivity increased with applied voltage, which indicates non-ohmic behavior. This is consistent with the literature, where it was observed that MWCNT nanocomposites exhibit tunneling conductive mechanisms, where a stronger applied electric field creates more conductive pathways through the material [52].

However, due to the large standard deviation caused by the variation in resistance measurements during this time period, statistical significance between means at different voltages and weight fractions was assessed using a two-way analysis of variance (ANOVA), followed by Tukey's multiple comparison test. The software GraphPad Prism 9.1.0 was used to carry out statistical analyses. The level of significance was set at $p < 0.05$. At 1 V and 2 V, the only significant comparisons ($p < 0.05$) were between 15 wt.% and 25 wt.% MWCNT. At all other voltages, 20 wt.% versus 25 wt.% MWCNT was significant. For the same wt.% value, the main ANOVA outcomes can be summarized as follows: no significance was determined between 1 V and 2 V, then between 6 V and 8 V, 6 V and 10 V, and 8 V and 10 V. Other comparisons between 1 V vs. 4 V, 6 V, 8 V and 10 V, between 2 V vs. 6 V, 8 V and 10 V, then between 4 V vs. 8 V and 10 V, were determined to be significant. Thus, this confirms the general increasing trend with MWCNT weight fraction and applied voltage.

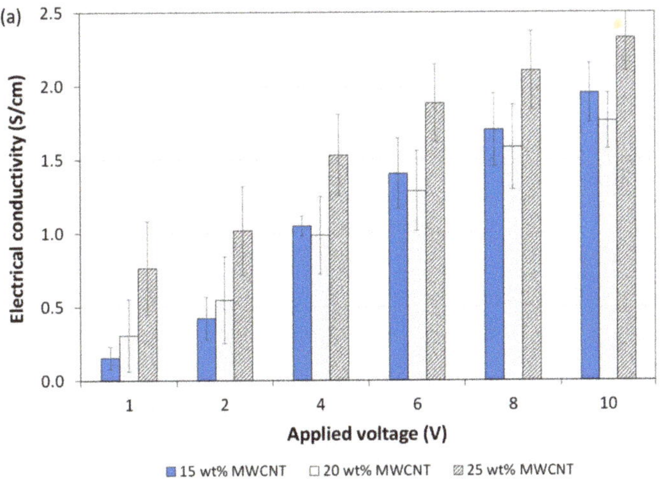

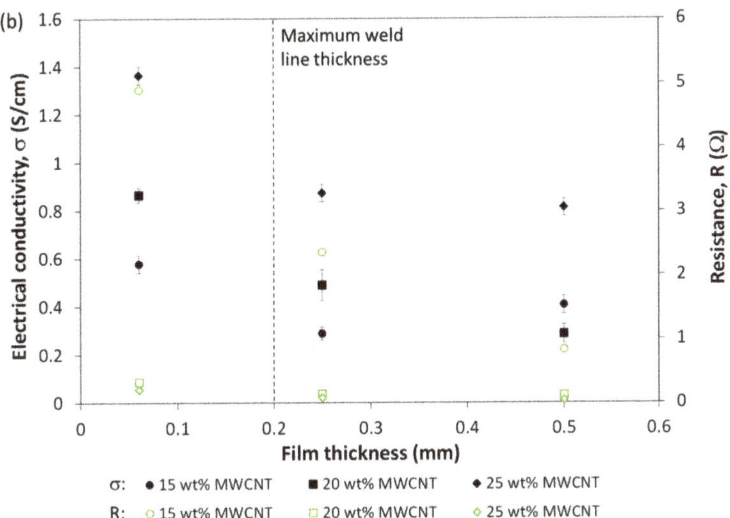

Figure 6. (**a**) Influence of applied voltage on average film electrical conductivity for 15 wt.%, 20 wt.% and 25 wt.% MWCNT; (**b**) influence of film thickness on electrical conductivity (filled markers) and resistance (unfilled markers) for 15 wt.%, 20 wt.% and 25 wt.% MWCNT. Representative data shown when voltage of 2 V was applied.

The effect of film thickness was assessed separately and a representative plot is shown in Figure 6b at an applied voltage of 2 V. Thickness is especially important with respect to the welding process as it is controlled through the vertical displacement (travel). In our study, a travel equal to 60% of the initial film thickness (0.50 mm) was used, meaning the final bond line thickness would be equal to 0.20 mm at most. For all MWCNT fractions, the general trend shows a decrease in conductivity with an increase in film thickness. This is opposite to what was observed in the literature for CNT/PDMS nanocomposites prepared with a centrifugal mixer [53]. However, the fabrication method used in our study, compression molding, can explain this behavior. As no additional mechanical mixing or solvent-based dissolution was employed, the CNT dispersion is likely not perfectly random and uniform across the thickness. This may affect the density of the conducting channels in the CNT network and in turn, the electrical conductivity.

Corresponding resistance values are reported in Figure 6b as well. As will be explained in Section 4.1.2, lower resistance leads to higher temperature increase through Joule heating. This indicates that higher MWCNT fractions or thicknesses would allow for reaching higher temperatures for the same applied voltage. Based on the trend shown in Figure 6b, a higher thickness at the weld line would be preferable for the disassembly procedure to insure the desired temperatures can be reached. Therefore, as was described in Section 2.4., the thickest nanocomposite films (0.50 mm) were used as energy directors for the ultrasonic welding process, leading to the highest bond line thickness.

4.1.2. Resistance Heating

Resistance heating follows Joule's Law, as shown by Equation (4), where P represents Watts of heating, I is the applied current (in A), V is the applied voltage (in V), and R is the electrical resistance (in Ohms) [54].

$$P = IV = \frac{V^2}{R}, \tag{4}$$

As Equation (4) shows, for any given voltage the amount of heating is controlled by the resistance of an object. Therefore, the lower the resistance, the more heat it will generate. Moreover, under a given voltage, thicker films would heat up more because, as shown by Equation (5) below, the resistance of an object will decrease with a larger cross-sectional area:

$$R = \rho \frac{L}{A}, \tag{5}$$

where ρ is the resistivity, L is the length (in mm), and A is the cross-sectional area (in mm^2). Examples of thermal profiles measured for nanocomposite films containing 15 wt.% and 20 wt.% MWCNT are shown in Figure 7. The maximum temperature generally increased with film thickness, CNT weight fraction and applied voltage (Equation (4)). In some cases, slight deviations from this trend were expected based on the large standard deviations seen in Figure 6a and resulting resistance values. At 15 wt.% MWCNT, maximum temperatures of 58.0 °C, 78.7 °C and 108.0 °C were obtained at 10 V, for 0.06 mm, 0.25 mm and 0.50 mm thickness, respectively. On the other hand, at 20 wt.% MWCNT, maximum temperatures of 80.2 °C, 96.1 °C and 116.0 °C were obtained at 10 V, for 0.06 mm, 0.25 mm and 0.50 mm thicknesses, respectively. Figure 8 shows a composite of the measured temperature profiles during the entire Joule heating experiment, when voltage was increased from 2 V up to 10 V for all film thicknesses. Maximum temperatures were obtained at 10 V with 101.6 °C, 102.8 °C and 120.2 °C for 0.06 mm, 0.25 mm and 0.50 mm film thicknesses, respectively.

For the GF/PP adherends and MWCNT/PP films used in this study, their melting temperature (T_m) was measured by differential scanning calorimetry (DSC) in a previous study [36]. The adherends' T_m was 150 °C, while the T_m of the MWCNT/PP films varied between 141 °C and 149 °C. For the purpose of disassembly, it is expected an interface temperature close to, or slightly above, this range of temperature should be reached through the energy director (MWCNT/PP film). As previously mentioned in Section 4.1.1, the bond line thickness is expected to be equal to 0.2 mm at most. Therefore, an applied voltage above 10V would be required for disassembly experiments, based on the trends observed in Figures 7 and 8.

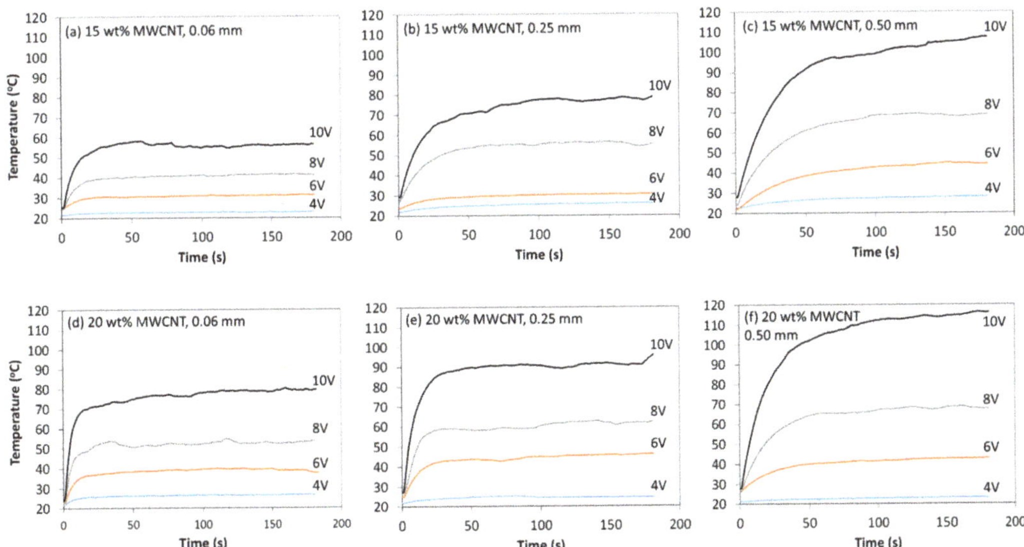

Figure 7. Temperature profiles of MWCNT/PP nanocomposite films at different input voltages and thicknesses: (**a–c**) 15 wt.% MWCNT, 0.06 mm, 0.25 mm and 0.50 mm thicknesses, respectively; and (**d–f**) 20 wt.% MWCNT, 0.06 mm, 0.25 mm and 0.50 mm thicknesses, respectively.

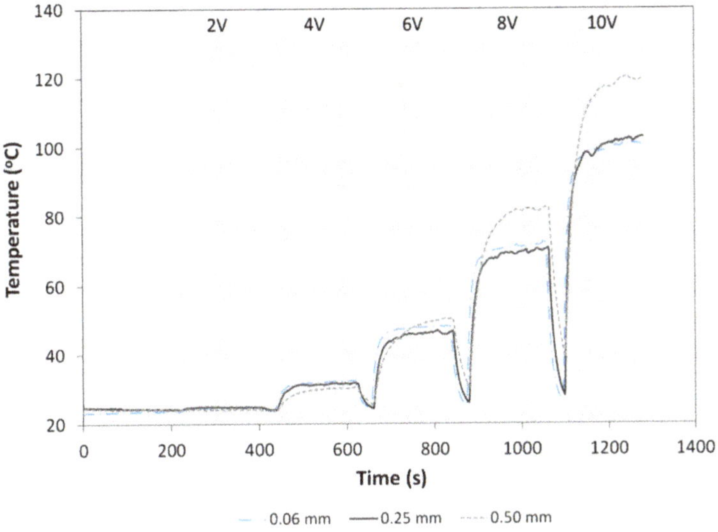

Figure 8. Representative temperature profiles measured for 25 wt.% MWCNT/PP films when voltages from 2 V to 10 V are applied for three minutes each.

4.2. Disassembly Study of Ultrasonically Welded Joints

4.2.1. Tensile Test Results

Feasibility of the disassembly procedure was assessed and quantified using a tensile testing machine. The lap shear strength (LSS) from load–displacement curves was calculated using the maximum load and the overlap area (25.4 mm × 127 mm). For each MWCNT weight fraction, disassembly was initiated at three adherend's surface temperatures: 110 °C, 130 °C and 150 °C. Figure 9 summarizes the calculated LSS for all cases and the range of LSS reduction when compared to room temperature tests. It is observed that, as surface temperature increased to 150 °C, the required strength for disassembly was reduced by up to 94%, corresponding to an applied tensile load below 250 N. The lowest surface temperature (110 °C) led to a considerable reduction in LSS, but as the interface temperature likely did not reach melting point, it is not as effective as higher temperatures. At 130 °C, there is a considerable difference between 15 wt.% MWCNT and 20 wt.% or 25 wt.% MWCNT films. There are potentially two causes for this behavior: (1) it was observed that an increase in MWCNT content could lead to lower toughness at the interface for welded joints [36,39,40]; (2) due to the slow crosshead speed during disassembly tests (1.3 mm/min), the temperature likely continued to increase at the interface, which may have been more significant at higher MWCNT loadings.

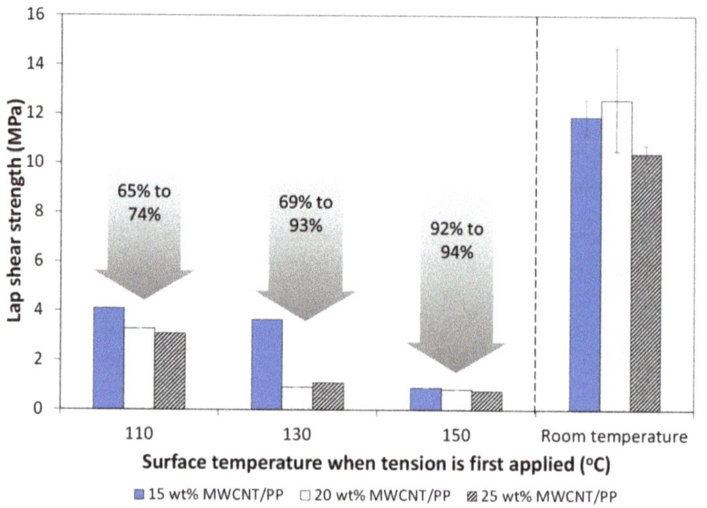

Figure 9. Comparison between lap shear strength of GF/PP welded joints during disassembly procedure when surface temperature reached 110 °C, 130 °C and 150 °C. Interface contained 15 wt.%, 20 wt.% and 25 wt.% MWCNT. Room temperature values are used as a reference, as reported in [36].

To further investigate the joints' behavior during disassembly, the shear stress and surface temperature curves were simultaneously plotted with respect to time, as seen in Figure 10. Representative curves are shown for all weight fractions on Figure 10a–c (15 wt.%, 20 wt.% and 25 wt.% MWCNT), at one surface temperature (110 °C, 130 °C and 150 °C). The heat up and disassembly phases are labeled to show the duration of each one. All tests were initiated after less than two minutes (120 s), with the fastest heat up phase for 25 wt.% MWCNT/PP films. Similarly, the disassembly phase generally lasted less than two minutes. As the applied load increased at the beginning of the disassembly phase, the surface temperature, and by extension, interface temperature, continued to increase as well because the contact at the weld line was not yet severed. However, after failure initiation (at the stress peak), the temperature slowly started to decrease as the integrity of the interface was compromised, leading to fewer conductive paths between MWCNTs. Since the applied voltage was kept constant throughout the disassembly procedure, the temperature consequently decreased. In some cases, as seen in Figure 10c, the disassembly phase displayed an inconsistent stress curve. One possible explanation is that, upon closer inspection of the specimens after disassembly, a small crack defect along the direction of the fibers was found in the adherends at the overlap. As all adherends were visually inspected after welding and no such defect was detected, it is reasonable to assume the crack was created during the disassembly process, likely explaining the inconsistent curves seen in Figure 10c. Another explanation is the reduced heating capacity resulting from failure initiation at the interface, and leading to cool down, as observed in the temperature curve. Cool down could contribute to an increase in stress during the process.

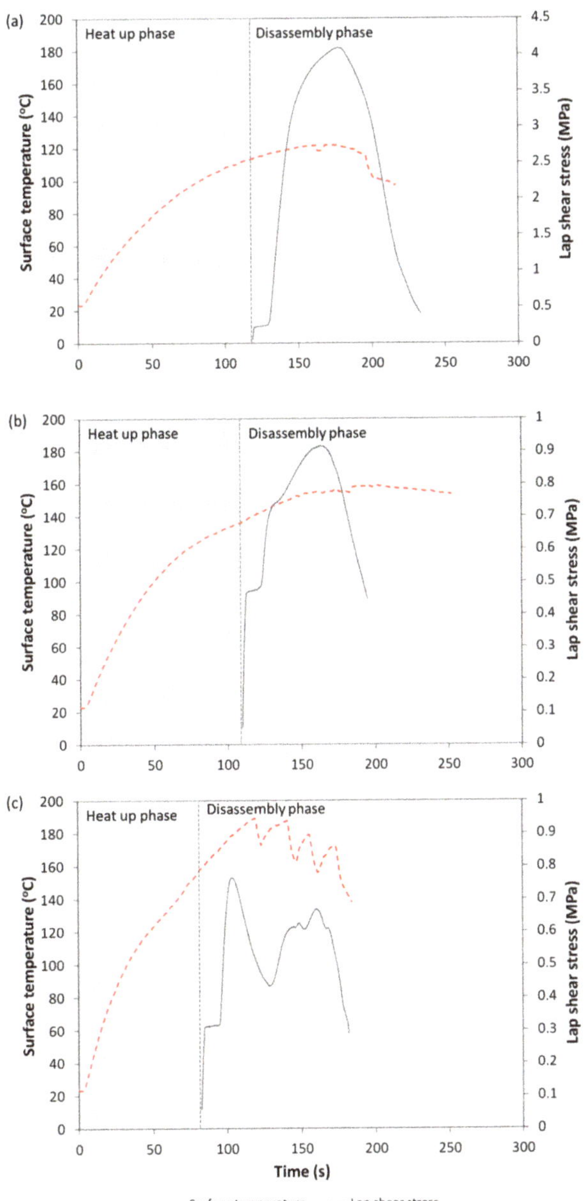

Figure 10. Representative lap shear stress curves (solid lines) of GF/PP welded joints during disassembly procedure with interface containing (**a**) 15 wt.% MWCNT at adherend's surface temperatures of 110 °C; (**b**) 20 wt.% MWCNT at adherend's surface temperatures of 130 °C; and (**c**) 25 wt.% MWCNT at adherend's surface temperatures of 150 °C. Corresponding surface temperature curves are shown as dashed lines.

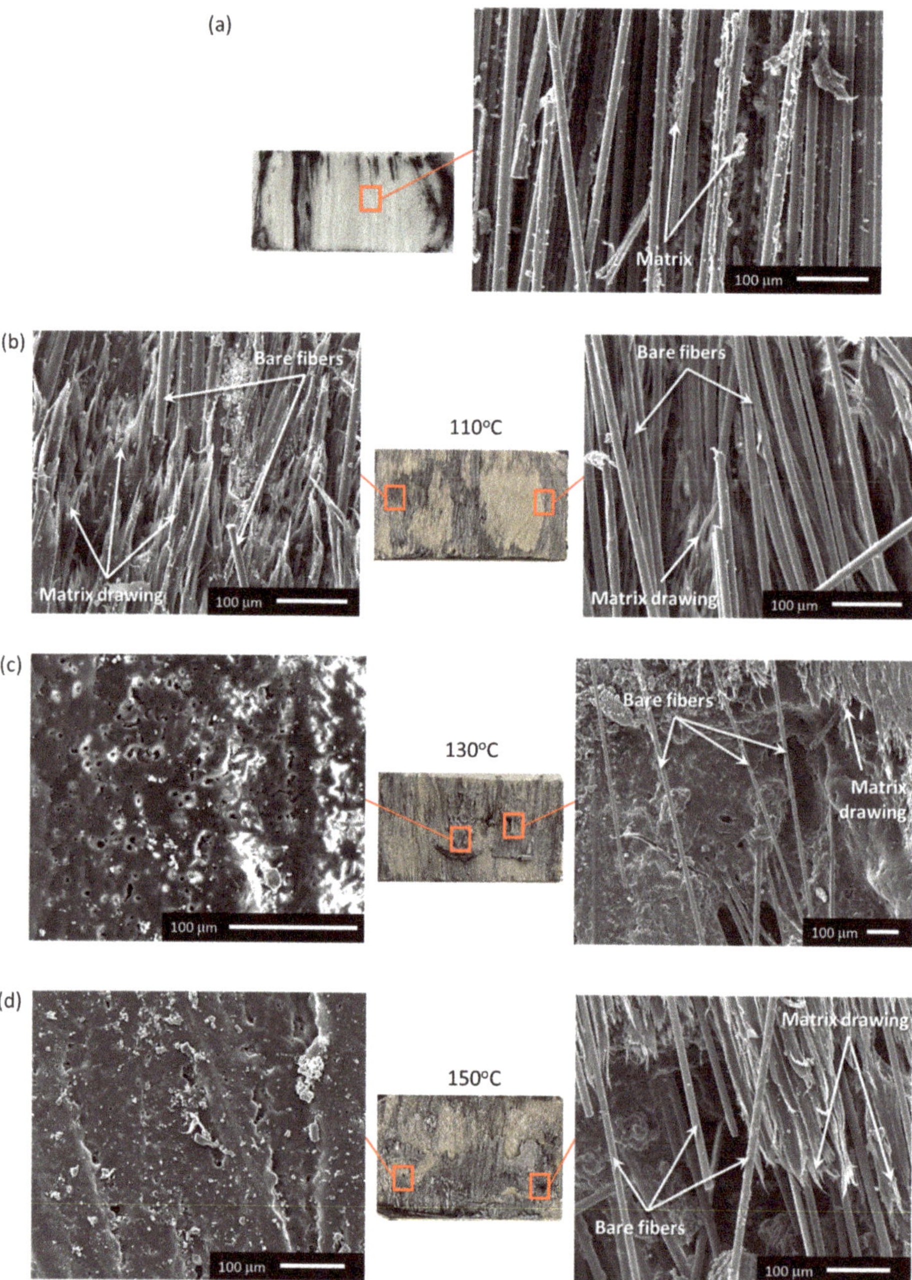

Figure 12. Representative fracture surfaces and SEM micrographs of samples welded with 20 wt.% MWCNT/PP films after disassembly process: (**a**) comparison with room temperature fracture surface images, reproduced and modified from [36] with permission; (**b**) disassembly at 110 °C; (**c**) disassembly at 130 °C; (**d**) disassembly at 150 °C. All scale bars are 100 μm.

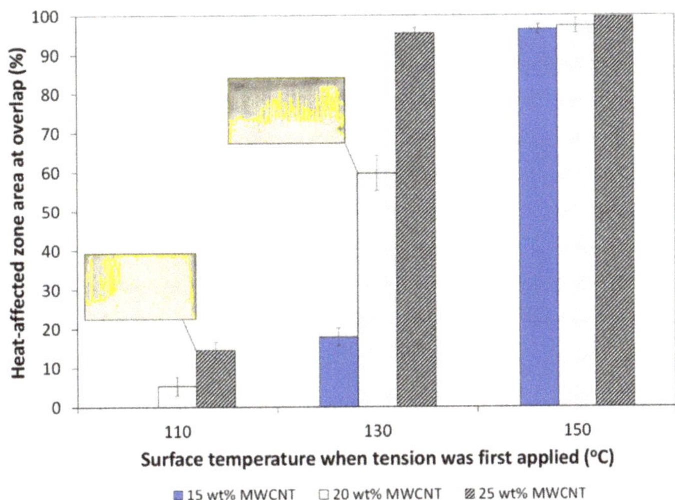

Figure 13. Estimated heat-affected zone area in GF/PP adherends after disassembly procedure at difference surface temperatures, based on adherends' surface color images presented in Figure 11. Two examples of delineated areas are shown in insets.

The HAZ reported in Figure 11, then quantified in Figure 13, is consistent with the temperature curves in Figure 10b,c, where a temperature above 150 °C was reached on the adherend's surface during the disassembly phase. Nonetheless, the experiments confirm resistance heating can facilitate disassembly of ultrasonically welded TPC joints through a manual process, especially at higher weight fractions (20 wt.% and 25 wt.% MWCNT) and surface temperatures (130 °C and 150 °C). Under the parameters investigated in this study, welds disassembled at a surface temperature of 130 °C with 20 wt.% MWCNT present the best balance between required shear stress and heat-affected zone.

Given the results presented in this study, a discussion on the limitations of this disassembly method and future work is warranted. It was demonstrated that resistance heating through an electrically conductive nanocomposite film at the welded interface can facilitate joint disassembly by lowering the required shear stress by more than 90%. However, as the process is relatively slow (<120 s heat-up phase) and the total interface/adherends thickness is low (<4 mm), the heat-affected zone extended through the thickness, mostly at higher temperatures (130 °C and 150 °C). Consequently, disassembly was not uniquely concentrated at the bond line where the MWCNT/PP film was placed, but affected the GF/PP adherends at the overlap as well. Thus, the method might be better suited for recycling at end-of-life or reuse of components by cutting off the damaged overlap section.

Finally, as the interface was structurally compromised during disassembly, it partially affected the efficiency of resistance heating. It is expected that a faster cross-head speed during disassembly, use of highest MWCNT weight fractions (such as 20 wt.% or 25 wt.%) and control of the applied voltage during the process could mitigate this limitation, as well as the extent of the HAZ. A faster cross-head speed would reduce the time between the beginning of the disassembly phase and the peak in the stress curves (Figure 10), as well as the overall duration of the disassembly phase. Therefore, the temperature when failure initiates and propagates at maximum stress would be lower, potentially limiting the HAZ in the adherends. Further, if failure were to occur at a faster rate, the interface might not have time to cool down due to lower heating efficiency. Some future research directions include (1) investigation of disassembly parameters (e.g., crosshead speed, voltage regulation through constant power output [39]); (2) use of thicker adherends to investigate HAZ; (3) healing of bond-line defects/damage through resistance heating.

5. Conclusions

In this work, it was demonstrated that resistance heating via an electrically conductive MWCNT/PP film at the welded interface facilitated ultrasonic joint disassembly of TPCs. Three MWCNT fractions were characterized for thermo-electrical behavior with applied voltages up to 10V. A maximum temperature of 120 °C was reached at the highest MWCNT loading and applied voltage. For disassembly experiments, tests were initiated when the surface temperature of the GF/PP adherend reached either 110 °C, 130 °C or 150 °C. The shear stress during disassembly decreased by at least 93% at the highest MWCNT weight fraction and surface temperature, compared to room temperature testing. Analysis of fracture surfaces after disassembly revealed the melting of both MWCNT films and the adherends' matrix at the overlap with significant matrix drawing and fiber–matrix debonding, effectively facilitating disassembly. At higher temperatures and MWCNT weight fractions, the heat-affected zone extended through the thickness of the adherends, owing to the low cross-head speed and the duration of the disassembly phase (<120 s) during which heat transfer occurred. In order to minimize the extent of the heat-affected zone area (<60%), while maximizing ease of assembly, a surface temperature of 130 °C with 20 wt.% MWCNT films would be recommended for the parameters investigated in this study.

Overall, this study confirmed the feasibility of this disassembly method for the first time in the literature. However, this might be better suited for recycling at end-of-life or reuse of components by cutting off the heat-affected overlap section. Moreover, as the interface was structurally compromised during disassembly, it affected the efficiency of resistance heating. Faster crosshead speeds during disassembly, the use of the highest MWCNT weight fractions, and control of the applied voltage during the process could, however, mitigate these limitations.

Author Contributions: Conceptualization, H.F. and G.P.; funding acquisition, G.P.; methodology, analysis and investigation, H.F. and W.L.; validation, H.F. and G.P.; writing—original draft preparation, H.F. and G.P.; writing—review and editing, G.P. and W.L.; visualization, H.F., G.P. and W.L.; supervision, G.P. All authors have read and agreed to the published version of the manuscript.

Funding: This research was funded by the Louisiana Board of Regents under the Research Competitiveness Subprogram (contract number LEQSF (2018-2022)-RD-A-05) with partial funding from the LaSPACE Research Enhancement Award (award number 002379) and the LSU Graduate School Economic Development Assistantship.

Data Availability Statement: Not applicable.

Acknowledgments: The authors would like to acknowledge Mark Brennan for his assistance with tensile tests.

Conflicts of Interest: The authors declare no conflict of interest. The funders had no role in the design of the study; in the collection, analyses, or interpretation of data; in the writing of the manuscript, or in the decision to publish the results.

References

1. Cousins, D.S.; Suzuki, Y.; Murray, R.E.; Samaniuk, J.R.; Stebner, A. Recycling glass fiber thermoplastic composites from wind turbine blades. *J. Clean. Prod.* **2019**, *209*, 1252–1263. [CrossRef]
2. Li, Y.; Liu, Z.; Shen, J.; Lee, T.H.; Banu, M.; Hu, S.J. Weld quality prediction in ultrasonic welding of carbon fiber composite based on an ultrasonic wave transmission model. *J. Manuf. Sci. Eng.* **2019**, *141*, 081010. [CrossRef]
3. Yousefpour, A.; Hojjati, M.; Immarigeon, J.-P. Fusion Bonding/Welding of Thermoplastic Composites. *J. Thermoplast. Compos. Mater.* **2004**, *17*, 303–341. [CrossRef]
4. Arnt, O.; van Ingen, J.W.; Buitenhuis, A. Development of a butt jointed thermoplastics stiffened shin concept. In Proceedings of the SAMPE Conference, Long Beach, CA, USA, 23–26 May 2011.
5. Agarwal, B.D.; Broutman, L.J.; Chandrashekhara, K. *Analysis and Performance of Fiber Composites*, 4th ed.; John Wiley & Sons, Inc.: Hoboken, NJ, USA, 2018.

6. Murray, R.E.; Penumadu, D.; Cousins, D.; Beach, R.; Snowberg, D.; Berry, D.; Suzuki, Y.; Stebner, A. Manufacturing and Flexural Characterization of Infusion-Reacted Thermoplastic Wind Turbine Blade Subcomponents. *Appl. Compos. Mater.* **2019**, *26*, 945–961. [CrossRef]

7. Reis, J.P.; de Moura, M.; Samborski, S. Thermoplastic Composites and Their Promising Applications in Joining and Repair Composites Structures: A Review. *Materials* **2020**, *13*, 5832. [CrossRef]

8. Engelschall, M.; Larsen, L.; Fischer, F.; Kupke, M. Robot-based continuous ultrasonic welding for automated production of aerospace structures. In Proceedings of the SAMPE Europe Conference, Nantes, France, 17–19 September 2019.

9. Jongbloed, B.; Teuwen, J.; Benedictus, R.; Villegas, I.F. On differences and similarities between static and continuous ultrasonic welding of thermoplastic composites. *Compos. Part B Eng.* **2020**, *203*, 108466. [CrossRef]

10. Bhudolia, S.K.; Gohel, G.; Leong, K.F.; Barsotti, R.J. Investigation on Ultrasonic Welding Attributes of Novel Carbon/Elium® Composites. *Materials* **2020**, *13*, 1117. [CrossRef] [PubMed]

11. Villegas, I.F.; Moser, L.; Yousefpour, A.; Mitschang, P.; Bersee, H.E. Process and performance evaluation of ultrasonic, induction and resistance welding of advanced thermoplastic composites. *J. Thermoplast. Compos. Mater.* **2013**, *26*, 1007–1024. [CrossRef]

12. Benatar, A.; Eswaran, R.V.; Nayar, S.K. Ultrasonic welding of thermoplastics in the near-field. *Polym. Eng. Sci.* **1989**, *29*, 1689–1698. [CrossRef]

13. Zhang, Z.; Xiaodong, W.; Yi, L.; Zhenqiang, Z.; Liding, W. Study on Heating Process of Ultrasonic Welding for Thermoplastics. *J. Thermoplast. Compos. Mater.* **2010**, *23*, 647–664. [CrossRef]

14. Bhudolia, S.K.; Gohel, G.; Kantipudi, J.; Leong, K.F.; Barsotti, R.J. Ultrasonic Welding of Novel Carbon/Elium® Thermoplastic Composites with Flat and Integrated Energy Directors: Lap Shear Characterisation and Fractographic Investigation. *Materials* **2020**, *13*, 1634. [CrossRef] [PubMed]

15. Tao, W.; Su, X.; Wang, H.; Zhang, Z.; Li, H.; Chen, J. Influence mechanism of welding time and energy director to the thermoplastic composite joints by ultrasonic welding. *J. Manuf. Process.* **2019**, *37*, 196–202. [CrossRef]

16. Palardy, G.; Villegas, I.F. On the effect of flat energy directors thickness on heat generation during ultrasonic welding of thermoplastic composites. *Compos. Interfaces* **2016**, *24*, 203–214. [CrossRef]

17. Villegas, I.F.; Bersee, H.E.N. Ultrasonic welding of advanced thermoplastic composites: An investigation on energy-directing surfaces. *Adv. Polym. Technol.* **2010**, *29*, 112–121. [CrossRef]

18. Yan, J.; Wang, X.; Li, R.; Xu, H.; Yang, S. The Effects of Energy Director Shape on Temperature Field During Ultrasonic Welding of Thermoplastic Composites. *Key Eng. Mater.* **2007**, *353–358*, 2007–2010. [CrossRef]

19. Bhudolia, S.K.; Gohel, G.; Leong, K.F.; Islam, A. Advances in Ultrasonic Welding of Thermoplastic Composites: A Review. *Materials* **2020**, *13*, 1284. [CrossRef] [PubMed]

20. Choudhury, M.R.; Debnath, K. Analysis of tensile failure load of single-lap green composite specimen welded by high-frequency ultrasonic vibration. *Mater. Today Proc.* **2020**, *28*, 739–744. [CrossRef]

21. Bhudolia, S.K.; Gohel, G.; Kah, F.L.; Barsotti, R.J. Fatigue response of ultrasonically welded carbon/Elium® thermoplastic composites. *Mater. Lett.* **2020**, *264*, 127362. [CrossRef]

22. Jongbloed, B.; Teuwen, J.; Palardy, G.; Villegas, I.F.; Benedictus, R. Continuous ultrasonic welding of thermoplastic composites: Enhancing the weld uniformity by changing the energy director. *J. Compos. Mater.* **2020**, *54*, 2023–2035. [CrossRef]

23. Goto, K.; Imai, K.; Arai, M.; Ishikawa, T. Shear and tensile joint strengths of carbon fiber-reinforced thermoplastics using ultrasonic welding. *Compos. Part A Appl. Sci. Manuf.* **2019**, *116*, 126–137. [CrossRef]

24. Kalyan Kumar, R.; Omkumar, M. Investigation and characterization of ultrasonically welded GF/PA6T composites. *Mater. Today Proc.* **2019**, *26*, 282–286. [CrossRef]

25. Tutunjian, S.; Dannemann, M.; Fischer, F.; Eroglu, O.; Modler, N. A Control Method for the Ultrasonic Spot Welding of Fiber-Reinforced Thermoplastic Laminates through the Weld-Power Time Derivative. *J. Manuf. Mater. Process.* **2018**, *3*, 1. [CrossRef]

26. Villegas, I.F.; Palardy, G. Ultrasonic welding of CF/PPS composites with integrated triangular energy directors: Melting, flow and weld strength development. *Compos. Interfaces* **2017**, *24*, 515–528. [CrossRef]

27. Tutunjian, S.; Dannemann, M.; Modler, N.; Kucher, M. A Numerical Analysis of the Temporal and Spatial Temperature Development during the Ultrasonic Spot Welding of Fibre-Reinforced Thermoplastics. *J. Manuf. Mater. Process.* **2020**, *4*, 30. [CrossRef]

28. Koutras, N.; Amirdine, J.; Boyard, N.; Villegas, I.F.; Benedictus, R. Characterisation of Crystallinity at the Interface of Ultrasonically Welded Carbon Fibre PPS Joints. *Compos. Part A Appl. Sci. Manuf.* **2019**, *125*, 105574. [CrossRef]

29. Li, Y.; Yu, B.; Wang, B.; Lee, T.H.; Banu, M. Online quality inspection of ultrasonic composite welding by combining artificial intelligence technologies with welding process signatures. *Mater. Des.* **2020**, *194*, 108912. [CrossRef]

30. Lionetto, F.; Dell'Anna, R.; Montagna, F.; Maffezzoli, A. Modeling of continuous ultrasonic impregnation and consolidation of thermoplastic matrix composites. *Compos. Part A Appl. Sci. Manuf.* **2016**, *82*, 119–129. [CrossRef]

31. Palardy, G.; Shi, H.; Levy, A.; Le Corre, S.; Villegas, I.F. A study on amplitude transmission in ultrasonic welding of thermoplastic composites. *Compos. Part A Appl. Sci. Manuf.* **2018**, *113*, 339–349. [CrossRef]

32. Levy, A.; Le Corre, S.; Poitou, A. Ultrasonic welding of thermoplastic composites: A numerical analysis at the mesoscopic scale relating processing parameters, flow of polymer and quality of adhesion. *Int. J. Mater. Form.* **2014**, *7*, 39–51. [CrossRef]

33. Levy, A.; Le Corre, S.; Villegas, I.F. Modeling of the heating phenomena in ultrasonic welding of thermoplastic composites with flat energy directors. *J. Mater. Process. Technol.* **2014**, *214*, 1361–1371. [CrossRef]

34. Suresh, K.S.; Rani, M.R.; Prakasan, K.; Rudramoorthy, R. Modeling of temperature distribution in ultrasonic welding of thermoplastics for various joint designs. *J. Mater. Process. Technol.* **2007**, *186*, 138–146. [CrossRef]
35. Wang, X.; Yan, J.; Li, R.; Yang, S. FEM Investigation of the Temperature Field of Energy Director During Ultrasonic Welding of PEEK Composites. *J. Thermoplast. Compos. Mater.* **2006**, *19*, 593–607. [CrossRef]
36. Li, W.; Frederick, H.; Palardy, G. Multifunctional films for thermoplastic composite joints: Ultrasonic welding and damage detection under tension loading. *Compos. Part A Appl. Sci. Manuf.* **2021**, *141*, 106221. [CrossRef]
37. Frederick, H.; Li, W.; Sands, W.; Tsai, E.; Palardy, G. Multifunctional films for fusion bonding and structural health monitoring of thermoplastic composite joints. In Proceedings of the SAMPE Conference, Charlotte, NC, USA, 6 July 2020.
38. Brassard, D.; Dubé, M.; Tavares, J.R. Modelling resistance welding of thermoplastic composites with a nanocomposite heating element. *J. Compos. Mater.* **2020**, *55*, 625–639. [CrossRef]
39. Brassard, D.; Dubé, M.; Tavares, J.R. Resistance welding of thermoplastic composites with a nanocomposite heating element. *Compos. Part B Eng.* **2019**, *165*, 779–784. [CrossRef]
40. Farahani, R.D.; Janier, M.; Dubé, M. Conductive films of silver nanoparticles as novel susceptors for induction welding of thermoplastic composites. *Nanotechnology* **2018**, *29*, 125701. [CrossRef] [PubMed]
41. Farahani, R.D.; Dubé, M. Novel Heating Elements for Induction Welding of Carbon Fiber/Polyphenylene Sulfide Thermoplastic Composites. *Adv. Eng. Mater.* **2017**, *19*, 1700294. [CrossRef]
42. Vattathurvalappil, S.H.; Hassan, S.F.; Haq, M. Healing potential of reversible adhesives in bonded joints. *Compos. Part B Eng.* **2020**, *200*, 108360. [CrossRef]
43. Vattathurvalappil, S.H.; Haq, M. Thermomechanical characterization of Nano-Fe_3O_4 reinforced thermoplastic adhesives and single lap-joints. *Compos. Part B Eng.* **2019**, *175*, 107162. [CrossRef]
44. Ashrafi, B.; Laqua, K.; Martinez-Rubi, Y.; Jakubinek, M.B.; Park, D.; Simard, B. Electrically responsive polyurethane-CNT sheets for sensing and heating. In Proceedings of the CANCOM 2017, Ottawa, ON, Canada, 17–20 July 2017.
45. Martinez-Rubi, Y.; Ashrafi, B.; Jakubinek, M.B.; Zou, S.; Laqua, K.; Barnes, M.; Simard, B. Fabrication of High Content Carbon Nanotube–Polyurethane Sheets with Tailorable Properties. *ACS Appl. Mater. Interfaces* **2017**, *9*, 30840–30849. [CrossRef]
46. Senders, F.; van Beurden, M.; Palardy, G.; Villegas, I.F. Zero-flow: A novel approach to continuous ultrasonic welding of CF/PPS thermoplastic composite plates. *Adv. Manuf. Polym. Compos. Sci.* **2016**, *2*, 83–92. [CrossRef]
47. Almasri, A.M. Predicting the thermal conductivity of polypropylene-multiwall carbon nanotubes using the Krenchel model. *Sci. Eng. Compos. Mater.* **2018**, *25*, 383–388. [CrossRef]
48. Kim, P.; Shi, L.; Majumdar, A.; McEuen, P.L. Thermal Transport Measurements of Individual Multiwalled Nanotubes. *Phys. Rev. Lett.* **2001**, *87*, 215502. [CrossRef] [PubMed]
49. Pan, Y.; Li, L.; Chan, S.H.; Zhao, J. Correlation between dispersion state and electrical conductivity of MWCNTs/PP composites prepared by melt blending. *Compos. Part A Appl. Sci. Manuf.* **2010**, *41*, 419–426. [CrossRef]
50. Ramírez-Herrera, C.A.; Pérez-González, J.; Solorza-Feria, O.; Romero-Partida, N.; Flores-Vela, A.; Cabañas-Moreno, J.G. Highest recorded electrical conductivity and microstructure in polypropylene–carbon nanotubes composites and the effect of carbon nanofibers addition. *Appl. Nanosci.* **2018**, *8*, 1221–1232. [CrossRef]
51. Gulrez, S.; Ali Mohsin, M.E.; Shaikh, H.; Anis, A.; Poulose, A.M.; Yadav, M.K.; Qua, E.H.P.; Al-Zahrani, S.M. A review on electrically conductive polypropylene and polyethylene. *Polym. Compos.* **2014**, *35*, 900–914. [CrossRef]
52. Nanni, F.; Mayoral, B.L.; Madau, F.; Montesperelli, G.; McNally, T. Effect of MWCNT alignment on mechanical and self-monitoring properties of extruded PET–MWCNT nanocomposites. *Compos. Sci. Technol.* **2012**, *72*, 1140–1146. [CrossRef]
53. Jang, S.-H.; Park, Y.-L. Carbon nanotube-reinforced smart composites for sensing freezing temperature and deicing by self-heating. *Nanomater. Nanotechnol.* **2018**, *8*. [CrossRef]
54. Stavrov, D.; Bersee, H.E.N. Resistance welding of thermoplastic composites-an overview. *Compos. Part A Appl. Sci. Manuf.* **2005**, *36*, 39–54. [CrossRef]
55. Etcheverry, M.; Barbosa, S.E. Glass Fiber Reinforced Polypropylene Mechanical Properties Enhancement by Adhesion Improvement. *Materials* **2012**, *5*, 1084–1113. [CrossRef]
56. Rohart, V.; Lebel, L.L.; Dubé, M. Effects of environmental conditions on the lap shear strength of resistance-welded carbon fibre/thermoplastic composite joints. *Compos. Part B Eng.* **2020**, *198*, 108239. [CrossRef]
57. Koutras, N.; Benedictus, R.; Villegas, I.F. Thermal effects on the performance of ultrasonically welded CF/PPS joints and its correlation to the degree of crystallinity at the weldline. *Compos. Part C Open Access* **2021**, *4*, 100093. [CrossRef]

Article

Building of Longitudinal Ultrasonic Assisted Turning System and Its Cutting Simulation Study on Bulk Metallic Glass

Shuo Shan [1,2], Pingfa Feng [2,3], Huiting Zha [1,2,*] and Feng Feng [1,2]

[1] Division of Advanced Manufacturing, Graduate School at Shenzhen, Tsinghua University, Shenzhen 518055, China; ss17@mails.tsinghua.edu.cn (S.S.); feng.feng@sz.tsinghua.edu.cn (F.F.)

[2] Lab of Intelligent Manufacturing and Precision Machining, Tsinghua Shenzhen International Graduate School, Tsinghua University, Shenzhen 518055, China; fengpf@mail.tsinghua.edu.cn

[3] State Key Laboratory of Tribology, Department of Mechanical Engineering, Tsinghua University, Beijing 100084, China

* Correspondence: zhahuiting123@sz.tsinghua.edu.cn

Received: 5 June 2020; Accepted: 13 July 2020; Published: 14 July 2020

Abstract: Bulk metallic glass (BMG) is a new kind of material which is made by rapid condensation of alloy. With excellent properties like high strength, high hardness, corrosion resistance, BMG is increasingly applied in mold manufacturing, weapon equipment and other fields. However, BMG is also one of hard-to-machine materials, which is arduous to be processed precisely and efficiently by the means of conventional cutting. Compared with conventional cutting, ultrasonic machining has a multitude of technological advantages such as reducing the cutting force, extending the tool life, etc. In ultrasonic machining, the ultrasonic electric signal is transformed into high frequency mechanical vibration on the tool, which changes the relationship between the tool and the workpiece in the process of machining. In this study, the longitudinal ultrasonic assisted turning (LUAT) system is established for processing BMG. Its resonant frequency and vibration characteristics are first simulated by modal analysis and harmonic response analysis, and then tested by displacement testing experiments, so that the suitable frequency and the amplitude for BMG turning can be selected and verified. On this basis, the two-dimensional turning finite element model is established to study the effect of ultrasonic vibration on cutting force under different cutting speeds. The research manifest that during the BMG turning, the assistance of longitudinal ultrasonic vibration can significantly reduce the average cutting force as well as the von Mises stress when the turning speed is below the critical turning speed. In addition, the tip of the tool contacts the workpiece discontinuously during cutting process which makes the instantaneous turning force in LUAT more periodic than that in conventional turning (CT).

Keywords: bulk metallic glass; ultrasonic assisted turning; finite element analysis; cutting force

---

1. Introduction

Bulk metallic glass (BMG), also called amorphous alloy or liquid metal because of its topologically disordered constituent atoms, is formed by the rapid condensation of alloy. In BMG, the absence of crystal defects such as grain boundaries and dislocations brings to a series of unique properties [1]. BMG has high strength, high hardness, low thermal expansion coefficient, low density and favorable corrosion resistance, which makes it a potential new engineering material [2–5]. However, these excellent properties of BMG on the other hand would lead to poor machinability. For example, during the cutting process of BMG, the cutting force is considerately large, which can damage the cutting tools, jeopardize the surface quality and hinders their wide scale application [6–8]. Karaguzel et al. [9]

carried out the orthogonal cutting experiment of BMG, and the cutting force empirical formula of BMG is fitted by the experimental results and process parameters. Fujita et al. [10] examined cutting characteristics of BMG by tuning with different parameters, presumed a slipping-off mechanism at planes of short intervals of BMG. When cutting Zr-based BMG in low speed, there will be adiabatic shear bands and cavities in serrated chips, while in cutting with high cutting speed or small rake angle tools, there will be sparks and chip oxidation in the material [11–14]. Under ambient temperature, when BMG breaks, the local temperature of the alloy exceeds the glass transition temperature or even the melting point [15]. It is of great significance to raise new processing methods or improve the processing technology of BMG for promoting its engineering application.

Ultrasonic assisted machining has been termed as one of hybrid process that uses ultrasonic vibration during the machining action. Ultrasonic electric signal is converted into high-frequency mechanical vibration by transducer. The mechanical vibration is amplified by horn and transmitted to the end of the tool. In the turning process, the relative velocity between the tool tip and the workpiece changes due to the effect of ultrasonic vibration. Lauwers et al. [16] investigated the ultrasonic assisted grinding of ceramics, showed that the vibration resulted in more craters and lowered process force, making it feasible to increase the productivity. Pujana et al. [17] carried out ultrasonic assisted drilling experiments on titanium alloy, the results manifested that the drilling force decreases about 20%, and the it decreases with the increase of amplitude. Sui et al. [18] examined high-speed ultrasonic cutting, and the results showed that the material removal rate of high-speed ultrasonic cutting can be improved by up to 90% compared with conventional cutting. Maurotto et al. [19] applied ultrasonic longitudinal vibration to titanium alloy cutting. The average cutting force could be reduced by 70% compared with that of conventional cutting. Nath et al. [20] conducted cutting experiments on tungsten carbide by using elliptical ultrasonic vibration machining, studied the influence of tool nose radius on the surface quality, and obtained a better machining surface at 0.6-mm tool nose radius. Due to the application of longitudinal ultrasonic vibration, when the instantaneous speed of high-frequency vibration of the tool tip is greater than the cutting speed, there is contact-separation phenomenon between the tool tip and the workpiece. This kind of discontinuous contact makes the extrusion of the tool to the workpiece become high-frequency hammering, which changes the cutting force, cutting temperature, chip morphology and surface morphology of the machining surface.

A host of researches showed that ultrasonic vibration can improve the processing effect of BMG. Luo et al. [21] applied ultrasonic assisted micro-punch to BMG. The certain areas of BMG are subjected to the high frequency vibration transmitted by the punch, which leads to viscous flow. BMG gradually become soft and a series of shapes and products can be successfully fabricated in relative low pressing force. Ma et al. [22] proposed ultrasonic assisted forming for BMG processing, which not only forms BMG rapidly so that the crystallization and oxidation can effectively be avoided, but also works in a large scale range. In ultrasonic assisted punching and ultrasonic assisted forming on BMG, the cutting force and cutting temperature are effectively restrained, and the surface quality and productivity are improved. Whereas at present, there are few studies in ultrasonic assisted turning on BMG, which is subject to further investigation.

In this study, a longitudinal ultrasonic assisted turning (LUAT) system is established for processing BMG at first, and its frequency and amplitude are obtained through modal analysis and harmonic response analysis. A displacement testing experiment is performed to verify the designed system. Then the critical turning speed of the device is calculated. On this basis, the turning model of BMG is established, and the turning simulation of BMG under different turning speeds is carried out by using the LUAT device mentioned above. Finally, the influence of longitudinal ultrasonic vibration on the turning force, von Mises stress and chip forming under different turning speeds during turning BMG is discussed.

2. Building of Longitudinal Ultrasonic Assisted Turning System

2.1. Design of the Piezoelectric Transducer and the Ladder Horn

When a force is applied to the piezoelectric crystal in a proper direction, its internal polarization state will change. An internal electric field will then be generated, and a bound charge proportional to the external force will appear on two surfaces of the piezoelectric crystal. On the contrary, when an external electric field is applied to the piezoelectric crystal, the electric energy can be converted into mechanical energy through the piezoelectric crystal due to the inverse piezoelectric effect. The piezoelectric transducer is made of this inverse piezoelectric effect. Piezoelectric transducer has no eddy loss, hysteresis loss and resistance loss, which brings it fairly simple structure and high stability. Moreover, piezoelectric transducer usually has high sensitivity as well as outstanding electromechanical coupling characteristics. All these superior properties make piezoelectric transducer the most popular transducer in ultrasonic machining industry. In this investigation, a piezoelectric transducer with Lead Zirconate Titanate (PZT) is designed to convert electrical signals into ultrasonic vibration. Parameters of PZT are presented Table 1.

Table 1. Parameters of piezoelectric ceramics.

Material	Poisson's Ratio (−)	Density (kg/m³)	Young's Modulus (GPa)
PZT-8	0.31	7600	77

The longitudinal sound velocity in PZT slices c_{PZT} can be calculated in Equation (1).

$$c_{PZT} = \sqrt{\frac{E}{\rho}\frac{1}{1-\sigma^2}} \tag{1}$$

where, E is young's modulus, ρ is density and σ is Poisson's ratio of PZT-8.

The wavelength in PZT slices λ_{PZT} could then calculated by $\lambda_{PZT} = c_{PZT}/f$. According to the design requirements of the longitudinal vibration transducer, the equivalent diameter of the PZT slices D_{PZT} should less than $\lambda_{PZT}/4$, considering the inner holes, set the D_{PZT} as 38 mm.

The output power of the PZT transducer is determined by the volume of PZT ceramic slice and its power capacity. In order to amplify the output power to satisfy the turning process, several pieces of PZT slices are piled. The number m_{PZT} could be acquired by Equation (2).

$$m_{PZT} = \frac{P}{P_d f \pi (D^2 - d^2)\frac{t}{4}} \tag{2}$$

where P_d is power capacity in W/cm³, f is vibration frequency and D, d and t represent outside diameter, inside diameter and thickness of ceramic slice, respectively.

The power capacity of PZT ceramic slice P_d is 2–3 W/cm³, the required output power P is set as 800 W, considering the slices need to be piled in pairs, set the number of slices as 4. Four pieces of PZT slices are piled with copper electrodes separating between them, the polarization of two adjacent slices are opposite so that the pile could output an overlapped amplitude. The whole transducer used in this investigation is shown in Figure 1.

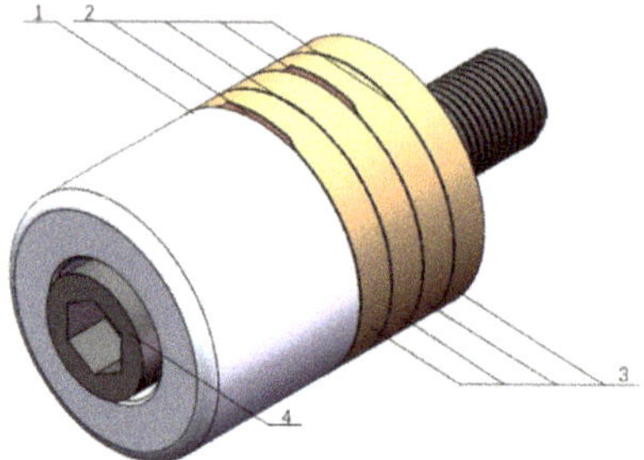

Figure 1. Structure diagram of the PZT transducer. 1. Back block; 2. electrodes; 3. PZT ceramics; 4. connect bolt.

The output vibration amplitude the piezoelectric transducer is much less than required. To amplify the amplitude and to install the turning tool, a two-stage ladder horn is designed. The back end of the horn is contact with the PZT slices, so the diameter of the back end D_1 is 38 mm. The front end is designed much smaller than the back end and the energy of ultrasound is concentrated on this comparative smaller area. Due to the energy gathering effect, the front end of the horn outputs the amplified amplitude. The diameter of the front end D_2 determines the amplification M. The less D_2 is, the higher M will be. However, the strength may decrease when D_2 is reduced because the horn can be seen as cantilever structure during turning process and the components of cutting force can lead to enormous stress concentration on abrupt change in section. To ensure the enough strength and satisfy the amplification, set D_2 as 10 mm. Thus, the area index is $N = D_1/D_2 = 3.8$.

Longitudinal sound wave will bounce back and forth between the back and front end, the standing-wave effect happens when the incident wave and the reflected wave overlap. At specific points, the amplitude can be decreased to zero. These points can be called as displacement node, which can be used to fix the horn to the machine tool. Likewise, the amplitude reaches its peak at particular points, where the turning tool should be installed. The displacement node and the peak point can be acquired though wave equation which is shown in Equation (3).

$$\frac{\partial^2 \xi}{\partial x^2} + \frac{1}{S} \times \frac{\partial S}{\partial x} \times \frac{\partial \xi}{\partial x} + k^2 \xi = 0 \tag{3}$$

where, $\xi = \xi(x)$ is particle displacement function, $S = S(x)$ is cross-section area function, $k = \omega / \sqrt{E/\rho}$ is circle wavenumber, ω is angular frequency, E is elastic modulus, ρ is the density of the horn.

In this investigation, the material of horn is 1045 steel, the parameters are listed in Table 2. Therefore, resonance length δ can be acquired as 258 mm, the length of the horn l is half of δ so that the front end located on the peak of the standing wave, outputting the largest amplitude.

Table 2. Parameters of 1045 steel.

Material	Longitudinal Wave Velocity (mm/s)	Density (kg/m³)	Elastic Modulus (GPa)
1045 steel	5.17×10^6	7890	21

The displacement node located at the middle of the horn, where a flange plate is designed to install the horn to the machine tool. The ladder horn is shown in Figure 2.

Figure 2. Two-stage ladder horn.

2.2. Modal and Harmonic Analysis of the LUAT System

Modal analysis is carried out to investigate the natural frequency and vibration shape of the turning system. The front face of the flange is set to be fixed to ground. After setting the contact region and meshing, 25 order resonant frequencies were calculated as shown un Figure 3. At the 12th, the 18th and the 19th modal the horn vibrates longitudinally, their resonant frequencies are 9493 Hz, 17,666 Hz and 18,315 Hz. Figure 4 describes these three modal shapes. As presented in figure, the 19th vibration mode is in positive agreement with requirements because it outputs the largest relative deformation, and its maximum vibration occurs at the front end of the horn, whereas the back end and the transducer only vibrates slightly. More detailed analysis will be examined in harmonic response analysis.

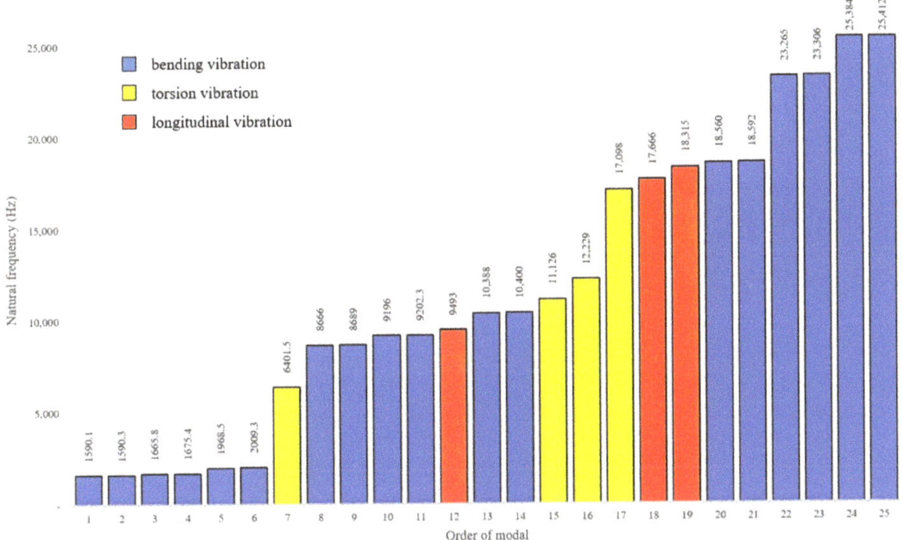

Figure 3. Natural frequency and vibration type of 25 orders of modals.

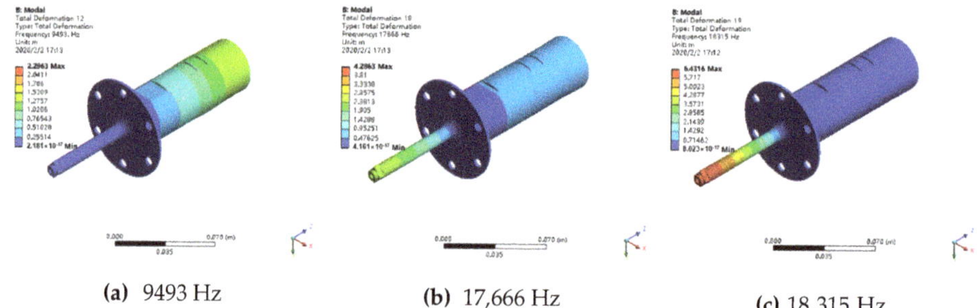

(a) 9493 Hz   **(b)** 17,666 Hz   **(c)** 18,315 Hz

Figure 4. Modal shape of the first three longitudinal vibrations. (**a**) 9493 Hz; (**b**) 17,666 Hz; (**c**) 18,315 Hz.

In the modal analysis, the vibration mode is relative value, which cannot reflect the true amplitude of the ultrasonic turning device at the tool. It is crucial to further analyze the harmonic response of the device on the basis of the 19th mode to obtain the amplitude at the tool tip. The harmonic response analysis is then conducted by workbench. Remote force is applied along the horn direction on its back end. The sweep range is set to between 17,000 Hz and 19,000 Hz (based on the 19th mode) with 100 solution intervals. Figure 5 shows the output displacement curve of the setting frequencies regarding the turning tool installed on the top of the horn. It could be observed that near to the frequency of 18,260 Hz, the structure is in the state of resonance, where the output displacement reaches the peak as 12 μm.

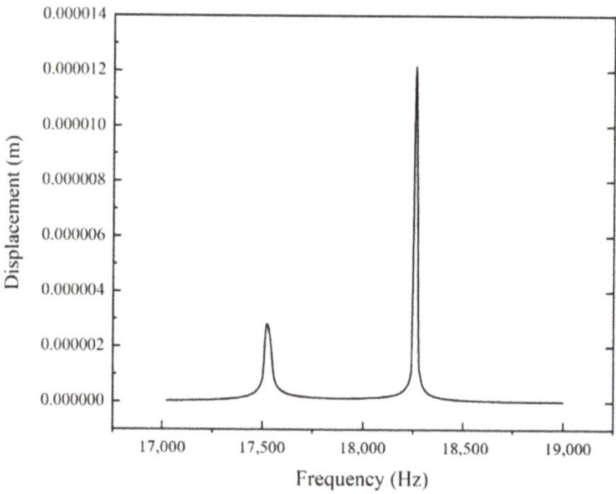

Figure 5. Displacement curve of the tool between 17,000 Hz and 19,000 Hz.

The designed LUAT system can output a considerable vibration with an amplitude up to 12 μm and a frequency of 18,260 Hz in the 19th mode, which is suitable for the turning process.

2.3. Vibration Testing Experiment of the LUAT System

A vibration testing experiment is conducted to verify the designed LUAT system, as shown in Figure 6. The LUAT system is activated by ultrasonic power supply BP6140. Tuning the frequency of the power supply so that the LUAT system reaches the resonant frequency, 18,265 Hz, which outputs the largest amplitude of vibration. Laser displacement sensor KEYENCE LK-H020 (KEYENCE CORPORATION, Osaka, Japan) is used to measure the displacement of the tool tip installed at the end of the ladder horn. Figure 6d depicts the displacement of the measured system, the amplitude can be calculated as around 12 μm. The result of the experiment verifies the structure design and the simulations of the LUAT system.

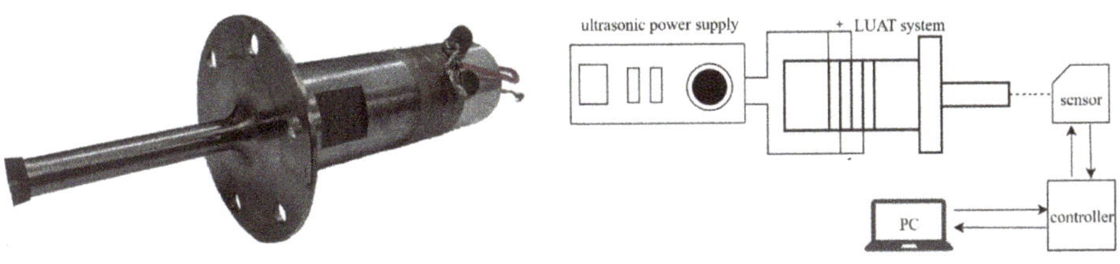

(a) Longitudinal ultrasonic assisted turning (LUAT) system

(b) Layout of displacement testing experiment

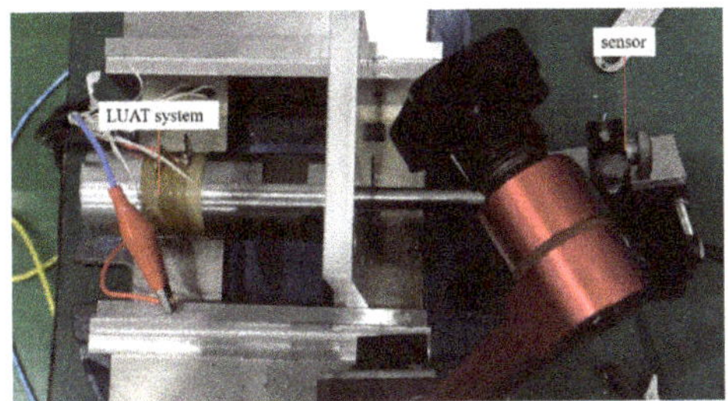

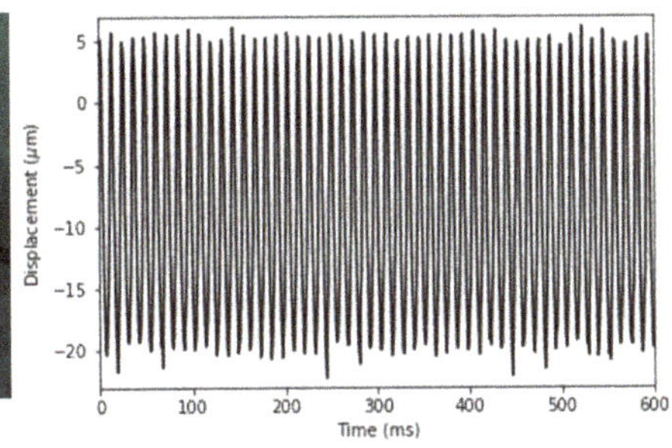

(c) Laser displacement sensor

(d) Displacement of LUAT

Figure 6. Displacement testing experiment. (**a**) longitudinal ultrasonic assisted turning (LUAT) system; (**b**) layout of displacement testing experiment; (**c**) laser displacement sensor; (**d**) displacement of LUAT.

3. Turning Simulation of Vit1 Bulk Metallic Glass

3.1. Finite Element Modeling of Vit1 BMG Turning Simulation

The accuracy of material model has a great influence on the success of simulation. In order to accurately reflect the nonlinear problem in the cutting process, it is necessary to comprehensively consider the stress–strain state of each point in the BMG under the complex strain state. The constitutive model of the material represents the stress–strain relationship of the material under load and describes it by mathematical expression. D–P model is a constitutive model of geotechnical materials. It not only considers the effect of intermediate principal stress on material yield, but also explains the experimental phenomena of "shear expansion effect" and "tension compression asymmetry" of metallic glass [8,23]. In this study, the D–P model is used as the constitutive model of vit1-metallic glass, and its main parameters [14,24] are shown in Table 3. Specifically, the D–P plasticity is defined by dilatancy angle, friction angle and flow–stress ratio, d-p hardening is defined by yield stress and absolute plastic strain.

Table 3. Related parameters of linear D–P model of BMG.

Material	Dilatancy Angle (°)	Friction Angle (°)	Flow–Stress Ratio (−)	Yield Stress (MPa)	Abs Plastic Strain (−)
Vit1	11.902	11.902	1	1194.62	0

The material parameters of vit1 BMG and turning tool are shown in Table 4. The constitutive model of metallic glass and the physical parameters of material are introduced into ABAQUS (Dassault Systèmes, Vélizy-Villacoublay Cedex, France), and the chip separation criterion is shear criterion.

Table 4. Physical parameters of Bulk metallic glass (BMG) and turning tool.

Material	Poisson's Ratio (–)	Density (kg/m³)	Young Modulus (GPa)
Vit1	0.36	6125	96,000
YG-8	0.22	14,700	640,000

Figure 7 shows the parameters and meshing of BMG turning simulation. YG-8 cemented carbide tool with 10° rake angle α and 7° relief angle β is chosen, and the nose radius r_ε is set as 0.2 mm. The longitudinal ultrasonic vibration is applied to the tool in the amplitude A paralleling to the relative turning motion. Due to the short cutting distance, the deformation of the turning tool could be ignored, and the tool is then set as a rigid body. In order to reduce the amount of calculation and ensure the accuracy of calculation, the side grid close to the cutting surface is relatively dense, and the side grid of the principle cutting surface is relatively sparse.

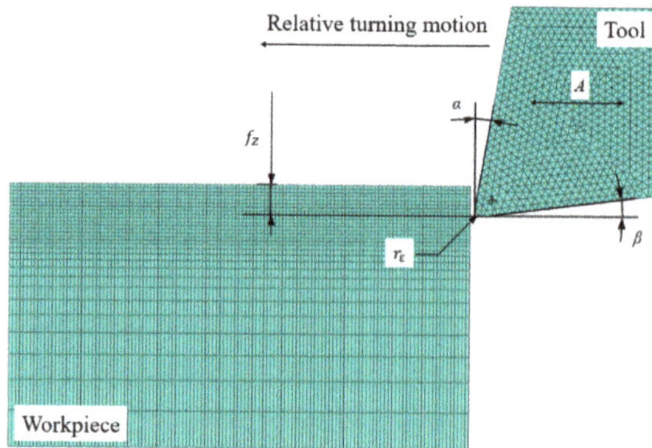

Figure 7. Parameters and meshing of BMG turning simulation.

3.2. Analysis of LUAT Process

In conventional turning (CT), the absolute speed of the tool tip in the peripheral direction of the workpiece is a fixed value. The introduction of ultrasonic vibration triggers the tool tip to produce high frequency reciprocating vibration in the peripheral direction. The average speed in a cycle $\overline{v_u}$ can be obtained from Equation (4):

$$\overline{v_u} = 2 \times A \times f \tag{4}$$

where, A and f represent the amplitude and the frequency of the tool tip vibration.

According to the modal analysis and harmonic response analysis, the suitable resonant frequency f is 18,260 Hz, and the displacement A is 12 μm. Consequently, the average speed $\overline{v_u}$ = 438.24 mm/s. The turning motion is equivalent to the fixed workpiece and the moving tool. At this time, the resultant velocity v can be regarded as the superposition of the workpiece velocity v_w and the instantaneous velocity of the tool v_u which comes from ultrasonic vibration.

$$v = v_w + v_u \tag{5}$$

Let v_u be cosine function and $v_u = a \cos \omega t$, then there is:

$$\overline{v_u} \times t = \int_0^t a \cos \omega t dt \tag{6}$$

Take t as half of the ultrasonic vibration period, then:

$$a = \frac{\overline{v_u} \times t \times \omega}{1 - \cos \omega t} \tag{7}$$

Plugging $\omega = 2\pi f$, it can be easily obtained that $a = 688.04$ mm/s. Therefor $v_u = 688.04 \cos 114{,}731t$, at this time the critical turning speed $v_t = \max v_u = 688.04$ mm/s.

The relationship between tool tip and workpiece is affected by the critical turning speed v_u of tool tip and workpiece velocity v_w. Specifically,

- When v_w is much larger than v_u, the direction of the closing speed v remains unchanged, and the size is first equal to v_w, which can be regarded as CT;
- When v_w is slightly larger than v_u, the tool tip keeps contact with the workpiece, the direction of closing speed remains unchanged and the size changes periodically;
- When v_w is smaller than the maximum value of v_u, the direction of closing speed changes periodically, and the contact separation state of tool tip and workpiece presents high frequency. At this time, the relationship between the tool and the workpiece is changed from the original tool extruding the workpiece to the high frequency tool hitting the workpiece.

In the range of 100 mm/s to 1800 mm/s, different v_w are selected to carry out the turning simulation experiment of BMG. The interval is 100 mm/s when v_w is below 600 mm/s and then rise to 200 mm/s when v_w is above 600 mm/s.

4. Results and Discussion

The turning simulation was conducted in different turning speeds by LUAT and CT, respectively in ABAQUS. Then the average cutting force, the instantaneous cutting force and the maximum von Mises stress are compared and analyzed.

4.1. Analysis of Average Cutting Force

The cutting force of the main turning direction is extracted from 50 sampling points when the tool is turned for 5 mm for analysis. The average cutting force of the LUAT and the CT at different v_w is shown in Figure 8.

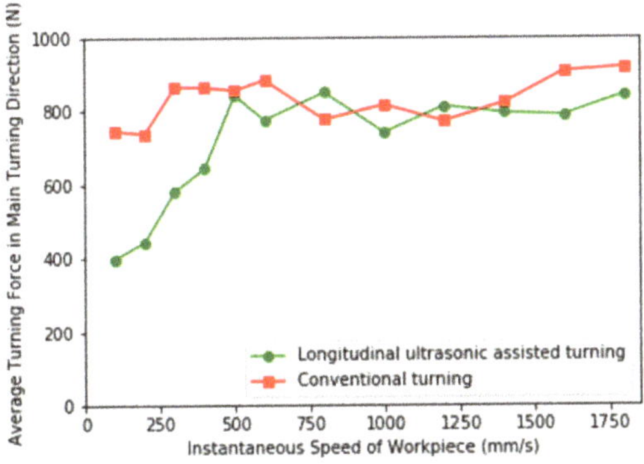

Figure 8. Average cutting force of LUAT and conventional turning (CT).

It can be observed from figure that when v_w is small, the average cutting force of LUAT is significantly smaller than that of CT, and the smaller v_w is, the more obvious the reduction of the average cutting force of longitudinal vibration is. When v_w is close to v_t, the effect of longitudinal ultrasonic vibration is gradually reduced. When v_w is greater than v_t, the turning force of LUAT has no obvious reduction compared with CT.

The reason for this phenomenon can be the contact-separation phenomenon between the tool tip and the workpiece. In the process of LUAT, the displacement where the tool nose cuts into the workpiece is extracted, and the displacement curve when the speed is 200 mm/s and 800 mm/s is shown in Figure 9. Where (a) is the displacement curve at 200 mm/s, when v_w is less than v_t, the direction of resultant velocity changes periodically under longitudinal vibration, and the tip of the tool contacts the workpiece discontinuously.

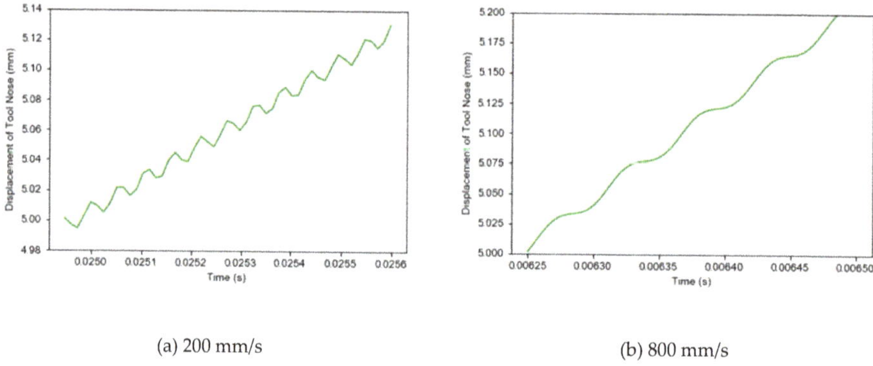

(a) 200 mm/s   (b) 800 mm/s

Figure 9. Displacement trajectory of the tool tip. (**a**) 200 mm/s; (**b**) 800 mm/s.

When $v_w < v_t$, the duty cycle $\alpha = t_1/T$, where t_1 represents the time when the tool movement has the same direction with the turning direction in a cycle and T is the cycle. From Equation (5), we can get,

$$t_1 = \frac{1}{\pi f} \cos^{-1}\left(-\frac{v_w}{v_t}\right) \tag{8}$$

Then,

$$\alpha = \frac{1}{\pi} \cos^{-1}\left(-\frac{v_w}{v_t}\right) \tag{9}$$

It can be seen from Equation (9) that α decreases with the decrease of v_w/v_t, and the decrease of instantaneous speed v_w of workpiece will increase the time when the resultant velocity is opposite to the main cutting speed in a cycle, and the time ratio of tool tip backward increases in displacement. The cutting force decreases significantly when the tool tip recedes, and then the average cutting force decreases.

4.2. Analysis of Instantaneous Cutting Force

In LUAT and CT, 80 sampling points are, respectively, extracted when the tool is turning 5 mm under the condition of $v_w = 100$ mm/s. The instantaneous cutting force in the main turning direction is compared in Figure 10.

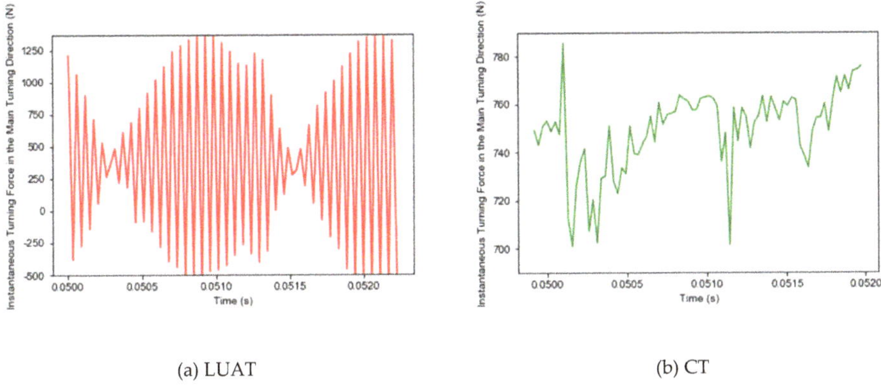

(a) LUAT   (b) CT

Figure 10. Instantaneous cutting force in main turning direction. (**a**) LUAT; (**b**) CT.

During the CT process, the tool extrudes the workpiece to deform, the stress and strain continue to accumulate until the workpiece material fails—this is how the chips generate. Conversely, as shown in Figure 11, due to the periodic contact separation between tool and workpiece, the stress form of workpiece changes from extrusion to high-frequency impact, and the peak value of instantaneous cutting force is significantly larger than that of CT.

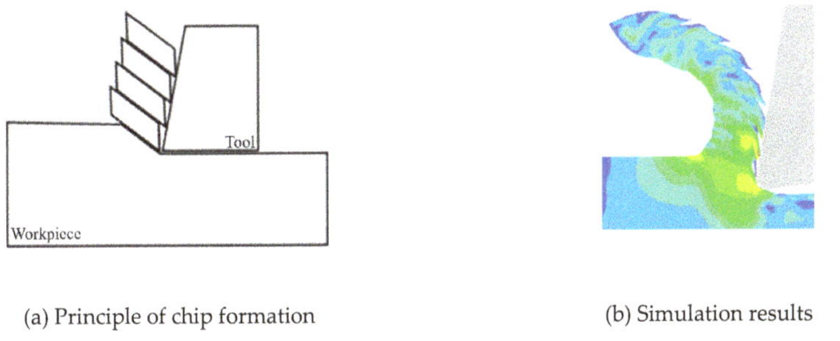

(a) Principle of chip formation   (b) Simulation results

Figure 11. Formation of BMG chips. (**a**) Principle of chip formation; (**b**) simulation results.

Furthermore, due to the contact-separation phenomenon, the periodic change of cutting force with chip breaking is more obvious. In CT, when the tool contacts the workpiece continuously, the BMG is still squeezed by the tool when the chip generates. In contrast, during the LUAT, the larger instantaneous cutting force and the impact of the tool under the contact separation state are more conducive to the failure of BMG, which promotes the generation of chips.

4.3. Internal Stress Analysis of Workpiece

The von Mises stress field obtained at different workpiece speeds from the simulation of LUAT and CT is shown in Figure 12. It can be noticed that in low-speed turning, the maximum von Mises stress of LUAT is significantly smaller than that of CT. With the increase of workpiece speed, the machining stress of LUAT increases gradually. In high-speed turning, the machining stress of LUAT is similar to that of CT. This is consistent with the observation from the average turning force above. Lower average turning force has an effect on von Mises stress in low speed LUAT. In addition, taking the contact-separate phenomenon into account, the relationship between the tool and the workpiece changes from continuously extrusion to discontinuously impact. The lower the turning speed is, the longer time the tool is separate to the workpiece in a single period, thus the materials in between can spring-back and release the stress. However, when the turning speed goes up, as the separated time decrease, the turning process becomes increasing continuously, hence the spring-back no longer happen. In short, the effect of longitudinal ultrasonic vibration only works at low turning speed when turning the BMG.

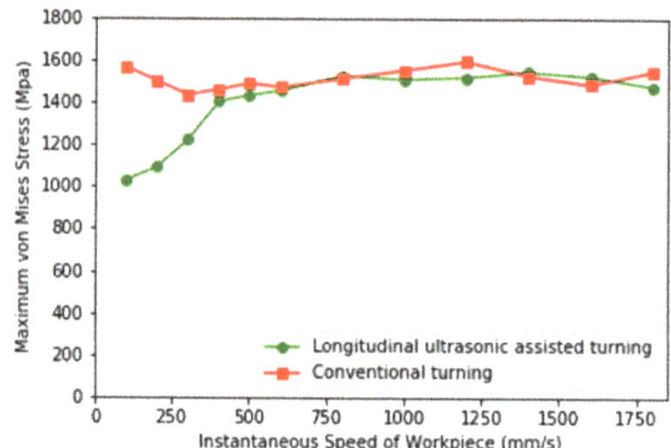

Figure 12. Maximum Mises stress of LUAT and CT.

Figure 13 depicts the von Mises stress nephogram of LUAT and CT at different turning speed in BMG turning. As depicted in figure, the maximum stress by LUAT in the first deformation area is obviously smaller than that by CT. Due to the phenomenon of discontinuous contact, the local deformation of the contact surface changes from extrusion to high-frequency impact, accordingly the stress can be released during separation. This results in the decrease of shear slip deformation and work hardening in the first deformation zone of the BMG. It can be easily observed that there is less residual stress in machined surface in LUAT, which represents less extrusion and friction during the turning process, given that the crystallization or even phase transition of BMG can be mitigated.

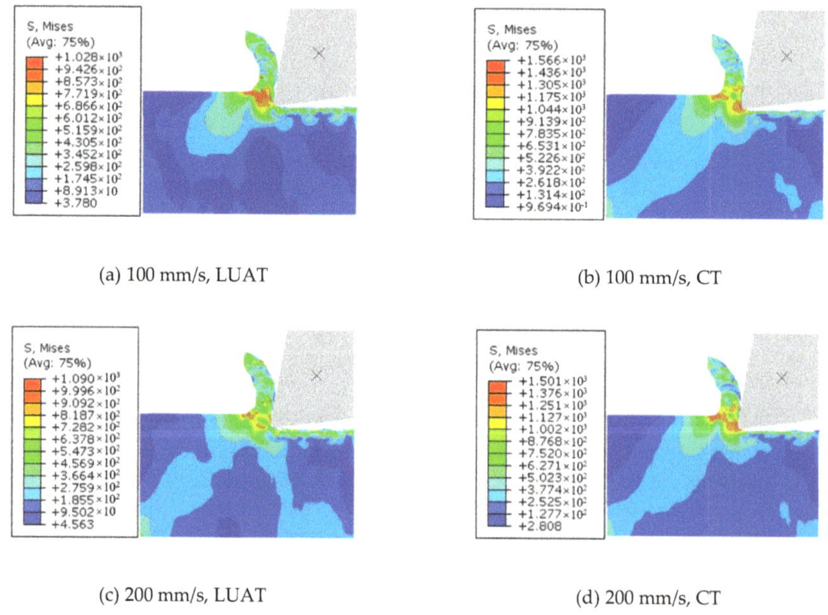

(a) 100 mm/s, LUAT

(b) 100 mm/s, CT

(c) 200 mm/s, LUAT

(d) 200 mm/s, CT

Figure 13. *Cont.*

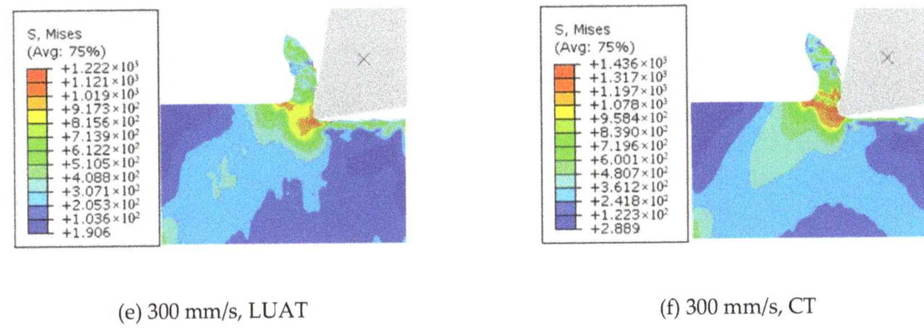

(e) 300 mm/s, LUAT   (f) 300 mm/s, CT

Figure 13. Maximum von Mises stress of LUAT and CT. (**a**) 100 mm/s, LUAT (**b**) 100 mm/s, CT; (**c**) 200 mm/s, LUAT (**d**) 200 mm/s, CT; (**e**) 300 mm/s, LUAT (**f**) 300 mm/s, CT.

Shapes and Von Mises stress in chips of LUAT and CT are depicted in Figure 14, respectively. During turning process, the BMG would be softened in the crack areas in chips, which can damage the tool and reduce the machined surface quality. In the case of LUAT, the residual stress in the chip is less than that in CT, which can improve the chip integrity. In addition, due to the promoting effect of chip formation, the chip of LUAT is more curved than that of CT, which is beneficial for releasing the residual stress, promoting chip removal during the turning process and improving the quality of machining.

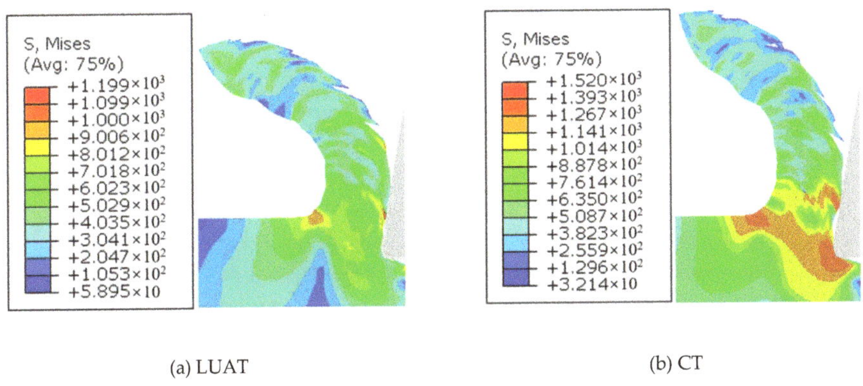

(a) LUAT   (b) CT

Figure 14. Shapes and von Mises stress in chips. (**a**) LUAT; (**b**) CT.

5. Conclusions

To improve the turning performance of BMG, first, a longitudinal ultrasonic vibration turning system was designed and then relative modal and harmonic response analysis was conducted. The analyses indicated that the vibration shape of the system met the design requirements at 19th mode. Through the experimental test, it turned out that the vibration amplitude of the system can reach 12 μm, which verified the design and the simulation. Therefore, the established LUAT system can realize ultrasonic assisted turning. Then the critical turning speed was derived, which proved a key factor of the effect of ultrasonic vibration and is determined by the parameters of the turning system. Finally turning simulations in different turning speeds were conducted on Vit1 BMG described by D–P constitutive model. For this work, the following conclusions can be drawn:

1. Simulation and theoretical analysis show that there is a critical turning speed in LUAT, which is determined by the longitudinal ultrasonic device. When the turning speed is lower than the critical turning speed, the longitudinal ultrasonic vibration can effectively reduce the average cutting force and von Mises stress of BMG cutting processing;

2. When the longitudinal ultrasonic vibration is applied in low speed turning, the tool and workpiece will have contact-separation phenomenon, which is conducive to the reduction of average cutting force and the periodic formation of chips;

3. During the turning process of BMG, the application of longitudinal ultrasonic vibration is rewarding to the formation of chips, so that the periodicity of instantaneous turning force is more obvious, which can help to form more regular chips and machined surface morphology;

4. LUAT can remarkably decrease the von Mises stress in both chips and machined surface, especially in the first deformation zone. This reduction is beneficial to improve the turning process of BMG.

Author Contributions: Conceptualization, S.S. and H.Z.; methodology, P.F.; validation, F.F.; investigation, S.S.; resources, F.F.; supervision, H.Z.; project administration, P.F. All authors have read and agreed to the published version of the manuscript.

Funding: This research was funded by Shenzhen Foundational Research Project (Subject Layout), Grant Number JCYJ20180508152128308 and Shenzhen Science and technology plan, Grant Number JSGG20180507181638381.

Conflicts of Interest: The authors declare no conflicts of interest.

References

1. Wang, W.H. The elastic properties, elastic models and elastic perspectives of metallic glasses. *Prog. Mater. Sci.* **2012**, *57*, 487–656. [CrossRef]

2. Zhu, P.-Z.; Qiu, C.; Fang, F.-Z.; Yuan, D.-D.; Shen, X.-C. Molecular dynamics simulations of nanometric cutting mechanisms of amorphous alloy. *Appl. Surf. Sci.* **2014**, *317*, 432–442. [CrossRef]

3. Hofmann, D.C. Bulk metallic glasses and their composites: A brief history of diverging fields. *J. Mater.* **2013**, *2013*, 517904. [CrossRef]

4. Chen, M.W. A brief overview of bulk metallic glasses. *NPG Asia Mater.* **2011**, *3*, 82–90. [CrossRef]

5. Liens, A.; Etiemble, A.; Rivory, P.; Balvay, S.; Pelletier, J.M.; Cardinal, S.; Fabregue, D.; Kato, H.; Steyer, P.; Munhoz, T.; et al. On the Potential of Bulk Metallic Glasses for Dental Implantology: Case Study on Ti40Zr10Cu36Pd14. *Materials* **2018**, *11*, 249. [CrossRef] [PubMed]

6. Basak, A.K.; Zhang, L.C. Deformation of Ti-Based Bulk Metallic Glass Under a Cutting Tip. *Tribol. Lett.* **2018**, *66*, 27. [CrossRef]

7. Browne, D.J.; Stratton, D.; Gilchrist, M.D.; Byrne, C.J. Bulk Metallic Glass Multiscale Tooling for Molding of Polymers with Micro to Nano Features: A Review. *Metall. Mater. Trans. A Phys. Metall. Mater. Sci.* **2013**, *44*, 2021–2030. [CrossRef]

8. Zhao, Y.; Lu, J.; Zhang, Y.; Wu, F.; Huo, D. Development of an analytical model based on Mohr–Coulomb criterion for cutting of metallic glasses. *Int. J. Mech. Sci.* **2016**, *106*, 168–175. [CrossRef]

9. Karaguzel, U.; Bakkal, M. Orthogonal cutting of Zr based bulk metallic glass by radial turning. In Proceedings of the International Conference On Materials Science, Metal and Manufacturing, Singapore, 12–13 December 2011.

10. Fujita, K.; Morishita, Y.; Nishiyama, N.; Kimura, H.; Inoue, A. Cutting Characteristics of Bulk Metallic Glass. *Mater. Trans.* **2005**, *46*, 2856–2863. [CrossRef]

11. Bakkal, M.; Liu, C.T.; Watkins, T.R.; Scattergood, R.O.; Shih, A.J. Oxidation and crystallization of Zr-based bulk metallic glass due to machining. *Intermetallics* **2004**, *12*, 195–204. [CrossRef]

12. Bakkal, M.; Shih, A.J.; McSpadden, S.B.; Liu, C.T.; Scattergood, R.O. Light emission, chip morphology, and burr formation in drilling the bulk metallic glass. *Int. J. Mach. Tools Manuf.* **2005**, *45*, 741–752. [CrossRef]

13. Bakkal, M.; Shih, A.J.; Scattergood, R.O.; Liu, C.T. Machining of a Zr–Ti–Al–Cu–Ni metallic glass. *Scr. Mater.* **2004**, *50*, 583–588. [CrossRef]

14. Bakkal, M.; Shih, A.J.; Scattergood, R.O. Chip formation, cutting forces, and tool wear in turning of Zr-based bulk metallic glass. *Int. J. Mach. Tools Manuf.* **2004**, *44*, 915–925. [CrossRef]

15. Liu, C.T.; Heatherly, L.; Easton, D.S.; Carmichael, C.A.; Schneibel, J.H.; Chen, C.H.; Wright, J.L.; Yoo, M.H.; Horton, J.A.; Inoue, A. Test environments and mechanical properties of Zr-base bulk amorphous alloys. *Metall. Mater. Trans. A Phys. Metall. Mater. Sci.* **1998**, *29*, 1811–1820. [CrossRef]

16. Lauwers, B.; Bleicher, F.; Ten Haaf, P.; Vanparys, M.; Bernreiter, J.; Jacobs, T.; Loenders, J.; Leuven, K. Investigation of the Process-Material Interaction in Ultrasonic Assisted Grinding of ZrO2 based Ceramic Materials. In Proceedings of the 4th CIRP International Conference on High Performance Cutting, Gifu, Japan, 24–26 October 2010.

17. Pujana, J.; Rivero, A.; Celaya, A.; de Lacalle, L.N.L. Analysis of ultrasonic-assisted drilling of Ti6Al4V. *Int. J. Mach. Tools Manuf.* **2009**, *49*, 500–508. [CrossRef]

18. Sui, H.; Zhang, X.Y.; Zhang, D.Y.; Jiang, X.G.; Wu, R.B.A. Feasibility study of high-speed ultrasonic vibration cutting titanium alloy. *J. Mater. Process. Technol.* **2017**, *247*, 111–120. [CrossRef]

19. Maurotto, A.; Muhammad, R.; Roy, A.; Silberschmidt, V. Enhanced ultrasonically assisted turning of a β-titanium alloy. *Ultrasonics* **2013**, *53*, 1242–1250. [CrossRef]

20. Nath, C.; Rahman, M.; Neo, K.S. A study on the effect of tool nose radius in ultrasonic elliptical vibration cutting of tungsten carbide. *J. Mater. Process. Technol.* **2009**, *209*, 5830–5836. [CrossRef]

21. Luo, F.; Sun, F.; Li, K.S.; Gong, F.; Liang, X.; Wu, X.Y.; Ma, J. Ultrasonic assisted micro-shear punching of amorphous alloy. *Mater. Res. Lett.* **2018**, *6*, 545–551. [CrossRef]

22. Ma, J.; Liang, X.; Wu, X.; Liu, Z.; Gong, F. Sub-second thermoplastic forming of bulk metallic glasses by ultrasonic beating. *Sci. Rep.* **2015**, *5*, 17844. [CrossRef]

23. Schuh, C.A.; Lund, A.C. Atomistic basis for the plastic yield criterion of metallic glass. *Nat. Mater.* **2003**, *2*, 449–452. [CrossRef] [PubMed]

24. Lu, J.; Ravichandran, G.; Johnson, W.L. Deformation behavior of the Zr41.2Ti13.8CU12.5Ni10Be22.5 bulk metallic glass over a wide range of strain-rates and temperatures. *Acta Mater.* **2003**, *51*, 3429–3443. [CrossRef]

Article

Effects of Ultrasonic Bending Vibration Introduced by an L-Shaped Ultrasonic Rod on the Microstructure and Properties of a 1060 Aluminum Alloy Strip Formed by Twin-Roll Casting

Chen Shi [1,2,3,*] **, Gaofeng Fan** [1] **, Xuqiang Mao** [3] **and Daheng Mao** [3]

[1] College of Mechanical and Electrical Engineering, Central South University, Changsha 410083, China; fangaofeng@csu.edu.cn

[2] State Key Laboratory of High Performance Complex Manufacturing, Central South University, Changsha 410083, China

[3] Light Alloy Research Institute, Central South University, Changsha 410083, China; maoxuqiang@csu.edu.cn (X.M.); mdh@csu.edu.cn (D.M.)

* Correspondence: shichen@csu.edu.cn; Tel.: +86-731-8887-7244

Received: 27 March 2020; Accepted: 23 April 2020; Published: 25 April 2020

Abstract: In order to achieve the industrial application of ultrasonic energy in the continuous casting and rolling production of aluminum alloy, a new type of L-shaped ultrasonic rod was used to introduce an ultrasonic bending vibration into the aluminum melt in the launder during the horizontal twin-roll continuous casting and rolling process of a 1060 aluminum alloy. The effects of the ultrasonic bending vibration on the microstructure and properties of the 1060 aluminum alloy cast rolling strip and its subsequent cold rolling strip were studied experimentally, and the effect of the ultrasonic-assisted refining with different amounts of Al-Ti-B refiner was explored. The results show that under the same addition amount of Al-Ti-B refiner, the ultrasonic bending vibration can refine the grains of the cast rolling strip, make the distribution of precipitates more uniform, reduce the slag inclusion defects, and improve the mechanical properties to a certain extent. The microstructure and properties of the ultrasonic cast rolling strip with 0.18 wt% Al-Ti-B refiner or 0.12 wt% Al-Ti-B refiner are better than those of the conventional cast rolling strip, but the microstructure and properties of the ultrasonic cast rolling strip with 0.09 wt% Al-Ti-B refiner are slightly worse than those of the conventional cast rolling strip. Moreover, after cold rolling, the effect of the ultrasonic bending vibration on the improvement of the microstructure and properties of the aluminum alloy strip is inherited. A comprehensive analysis shows that the use of ultrasonic energy in this paper cannot completely replace the effect of the Al-Ti-B refiner, but it can reduce the addition amount of the Al-Ti-B refiner by 1/3.

Keywords: ultrasonic bending vibration; 1060 aluminum alloy; twin-roll casting; microstructure; mechanical properties

---

1. Introduction

Aluminum alloy is the most widely used non-ferrous metal structure material in the metallurgy, chemical, construction, transportation, aerospace, and weapons industries [1,2]. With the application of aluminum alloys in the high-tech field, stricter requirements are imposed on the structure and properties of aluminum alloys [3,4]. At present, grain refinement by adding an Al-Ti-B refiner is an effective way to improve the performance of aluminum alloy strips in casting and rolling production [5–7]. A large number of researchers have studied and improved the Al-Ti-B refiner to improve its refining efficiency [8–11]. However, the addition of the Al-Ti-B refiner will generate several undesirable by-products, including the formation of particle aggregates, local defects, and impurities; in addition, the refiner when used as a consumable also increases the cost of the casting and rolling production.

In recent years, the use of additional physical fields such as ultrasonic vibration or pulsed magneto-oscillation (PMO) to replace the refiner used to refine the grains has aroused interest among researchers [12–16]. Xia et al. [17] used ultrasonic vibration to treat a 3003 aluminum alloy in continuous casting and rolling and found that the effect of ultrasonic treatment was better than that of adding Al-Ti-B refiner and that the refiner could be completely replaced, reducing the production cost and improving the material properties of the strip. Shi et al. [18] applied ultrasonic vibration to the process of 8011 aluminum alloy twin roll casting and rolling, and the research results showed that through the effect of ultrasonic vibration, the grains of the strip were made smaller and the mechanical properties were improved to a certain extent. Xu et al. [19] reduced the amount of Al-Ti-B by applying a electromagnetic field to the cast rolling of the 1100 aluminum alloy. It was found that the refining effect of adding 0.1 wt% Al-Ti-B refiner under the electromagnetic field roll-casting conditions was better than that of adding 0.4 wt% Al-Ti-B refiner under the conventional roll-casting conditions.

The key to the realization of ultrasonic-assisted metal solidification forming is the introduction of the ultrasonic wave. High-energy ultrasound can cause a cavitation effect and acoustic flow effect in the melt. With the increase in ultrasonic power, the effects will be significantly enhanced [20]. Due to these effects, the homogeneity of solute elements and fluidity of the melt can be enhanced, as well as improving the grain refinement and melt degassing efficiency [21–24]. At present, most researchers use a method of directly introducing ultrasonic waves in the upper part of the melt to treat the metal melt. The ultrasonic transducer is subjected to direct thermal radiation and a thermal shock from the molten metal, which often causes the transducer to be detuned or even damaged during the casting process. In addition, the titanium alloy ultrasonic radiator in contact with the melt is easily eroded [25–27]. These have made it difficult to achieve long-term ultrasonic effects in metal melts.

In recent years, our research group has developed a new type of L-shaped ultrasonic rod which uses a nano ceramic radiator to avoid the ultrasonic transducer being directly affected by high-temperature heat radiation and resist the erosion of metal melt, which can meet the requirements of long-term continuous work and is conducive to industrial application. Shi et al. [28] applied the L-shaped ultrasonic device to the solidification process of a large 2A14 aluminum alloy ingot (φ830 mm × 6000 mm). The study showed that it can significantly refine the grains of the large ingot, effectively decrease the degree of solute segregation, and improve its mechanical properties. The mechanical vibration generated by the ultrasonic transducer was conducted by an L-shaped ultrasonic rod, which formed an ultrasonic bending vibration at the head of the ceramic tool and led to metal melt. Based on the industrial test, this paper studies the influence of ultrasonic bending vibrations on the microstructure and properties of a 1060 aluminum alloy strip in the casting and rolling process and explores the ultrasonic-assisted refining effect of Al-Ti-B refiner in different usage conditions.

2. Material and Methods

2.1. Experimental Equipment

A new self-made L-shaped ultrasonic rod was used in the experiment, as shown in Figure 1. The L-shaped ultrasonic rod was composed of a transducer, a first-level horn that transmitted longitudinal mechanical vibrations in the horizontal direction, a second-level horn that transmitted bending vibrations in the vertical direction, and a ceramic radiator. The material used for the horn was a TC4 titanium alloy, and the material used for the ceramic radiator in direct contact with the metal melt was nano-silicon nitride ceramic. The working frequency of the ultrasonic wave was 21 ± 0.2 kHz and the power range was 0–1000 W.

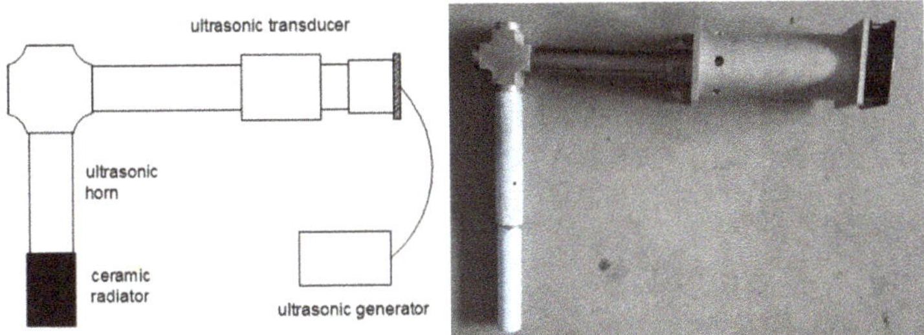

Figure 1. L-Shaped ultrasonic rod device.

2.2. Experimental Material

The experimental material was the 1060 aluminum alloy with the chemical composition listed in Table 1.

Table 1. Chemical compositions of the 1060 aluminium alloy (wt%).

Si	Cu	Mg	Zn	Mn	Ti	V	Fe	Al
≤0.25	≤0.05	≤0.05	≤0.05	≤0.05	≤0.03	≤0.05	0–0.4	Bal.

2.3. Experimental Procedures

The experiment was performed on a horizontal two-roll continuous cast rolling unit. Four groups of continuous casting and rolling experiments were carried out on the 1060 aluminum alloy: (1) adding 0.18 wt% Al-Ti-B refiner, conventional casting and rolling; (2) adding 0.18 wt% Al-Ti-B refiner and applying an ultrasonic bending vibration; (3) adding 0.12 wt% Al-Ti-B refiner and applying an ultrasonic bending vibration; (4) adding 0.09 wt% Al-Ti-B refiner and applying an ultrasonic bending vibration. The process of the 1060 aluminum alloy ultrasonic-assisted twin-roll continuous casting and rolling is shown in Figure 2.

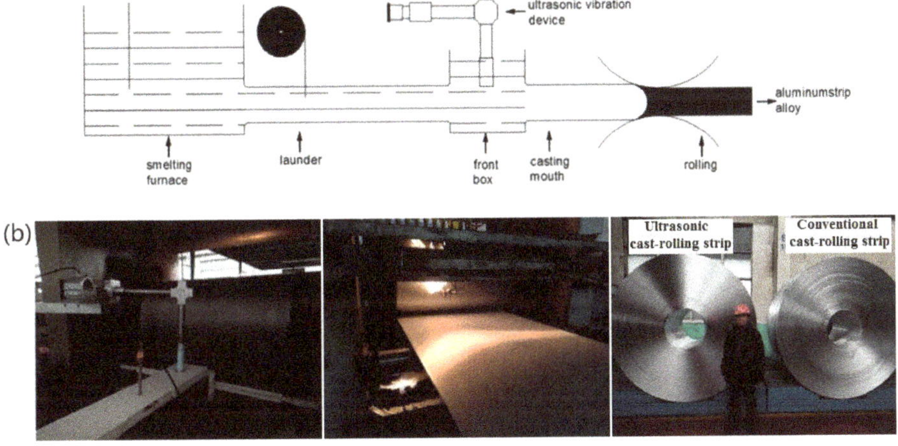

Figure 2. Schematic (**a**) and field picture (**b**) of the continuous casting and rolling of the 1060 aluminum alloy.

The technological parameters of the twin-roll casting unit were as follows: the roll gap was 5.7 mm, the twin roll casting speed was 0.85 m/min, the pouring temperature of the aluminum melt measured by the thermocouple was 695 °C, the temperature of the cooling water inlet in the roll was 34 °C, the water outlet was 41 °C, and the water pressure was 0.9 MPa. The parameters of the ultrasonic equipment were as follows: the power of ultrasonic wave was 800 W, the frequency was 21 ± 0.2 kHz, and the ceramic radiator was inserted into the aluminum melt surface in the launder at about 50 mm.

2.4. Test Method

2.4.1. Microstructure Analysis of the Cast Rolling Strip Samples

The $20 \times 20 \times 6$ mm samples were cut at a distance of 20 mm from the edge of the cast rolling strips. The samples were ground and polished in the MP-2B polishing machine and the morphology, size, and distribution of the precipitated phases were observed using a Phenom fully automatic scanning electron microscopy (Phenom-world BV, The Netherlands). Then, the electrolyte (19 mL water, 38 mL concentrated sulfuric acid, 43 mL phosphoric acid) was prepared for anodic coating. The current was controlled at 0.2 A, the temperature was controlled at 20 °C~30 °C, and the surface was observed every 30 s. After the grains appeared, the pictures of the grains on different sides of the cast rolling strip samples were obtained by a metallographic microscope (OLYMPUS DSX500 metallurgical microscope (Olympus, Japan)).

2.4.2. Microstructure Analysis of the Cold Rolling Strip Samples

In order to investigate the evolution of the microstructure and properties of the cast rolling strip after cold rolling, 4 groups of cold rolling strips with 0.3 mm thickness were obtained after the cold rolling of 4 groups of cast rolling strips. The 10 mm × 20 mm rectangular samples were cut in the middle of the cold rolling strips. After the electrolytic (90 mL of alcohol, 5 mL of water, 7 mL of perchloric acid) was prepared for electrolytic polishing, the morphology, size, and distribution of the precipitated phase were observed using a Phenom fully automatic scanning electron microscope (Phenom-world BV, The Netherlands). Then, an electrolytic (40 mL of water, 1 mL of HF) was prepared for anode coating, and grain pictures were obtained through a metallographic microscope (OLYMPUS DSX500 metallurgical microscope (Olympus, Japan)).

2.4.3. Test Mechanical Properties

Refer to the ASTME8M-04 metal material tensile test method for the tensile property test. Standard tensile samples were cut along the 0° direction of the strip (casting and rolling direction, RD), 45°, and 90° (transverse casting and rolling direction, TD). The tensile properties of the samples were tested at room temperature with the MTS 810 universal material testing machine (USA). The strain rate was 2 mm/min.

3. Results and Discussions

3.1. Effect of the Ultrasonic Bending Vibration on the Grain Microstructure of the 1060 Aluminum Alloy Cast Rolling Strip and Subsequent Cold Rolling Strip

The microstructure of the 1060 aluminum alloy after continuous casting and rolling is shown in Figure 3. It can be seen from the figure that the grains on the upper surface of the cast rolling strips are obviously elongated and have become fibrous; on the longitudinal section of the cast rolling strips, the grains are also fibrous. In comparison, it was found that the fibrous grains of the ultrasonic cast rolling strip tested with the 0.18 wt% Al-Ti-B refiner and the ultrasonic cast rolling strip tested with the 0.12 wt% Al-Ti-B refiner were finer, with average grain sizes of approximately 39.6 μm and 34.5 μm, respectively. The average grain size of the conventional cast rolling strip tested with the 0.18 wt% Al-Ti-B refiner was about 50.2 μm, and the grain size of the ultrasonic cast rolling strip tested with the

0.09 wt% Al-Ti-B was somewhat coarse and unevenly distributed, and the average grain size was about 54.0 μm. On the cross section of the cast rolling strips, the grain size of the ultrasonic cast rolling strip tested with the 0.18 wt% Al-Ti-B refiner and the ultrasonic cast rolling strip tested with the 0.12 wt% Al-Ti-B refiner were fine and uniform, and the average grain size was about 29.5 μm and 28.6 μm, respectively. Meanwhile, the average grain size of the conventional cast rolling strip with the 0.18 wt% Al-Ti-B refiner was about 33.4 μm, and that of the ultrasonic cast rolling strip with the 0.09 wt% Al-Ti-B refiner was coarser and unevenly distributed, with an average grain size of about 35.1 μm.

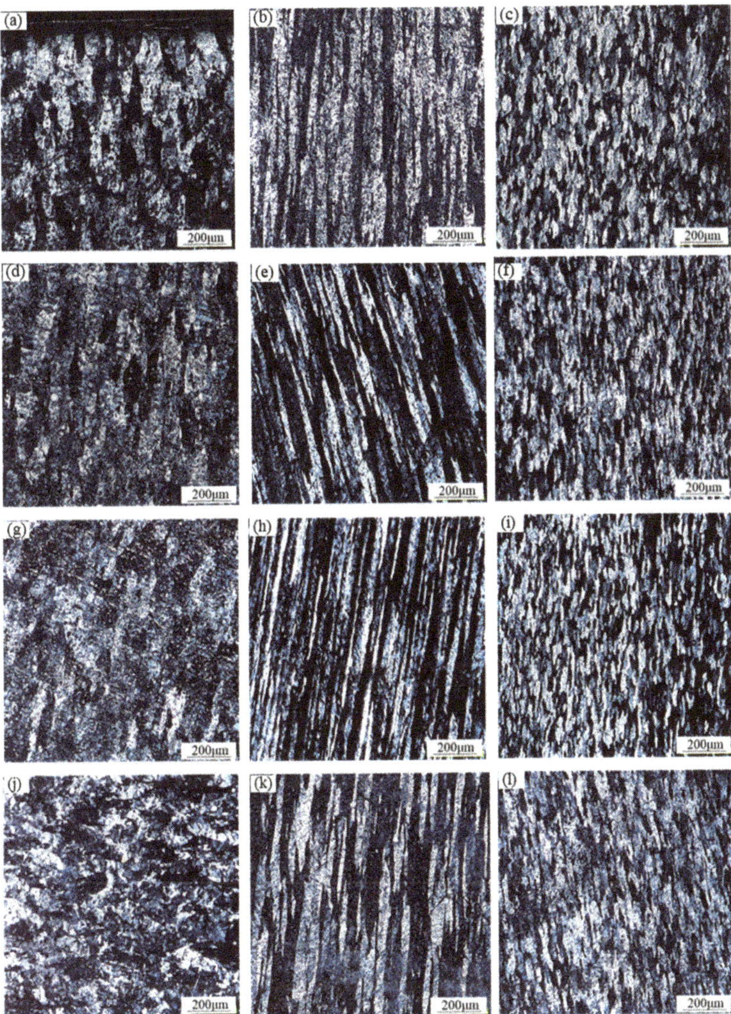

Figure 3. Grain microstructure of the 1060 aluminum alloy cast rolling strips: (**a–c**) the top surface, longitudinal section, cross section of the strip with added 0.18 wt% Al-Ti-B refiner and applied conventional treatment; (**d–f**) the top surface, longitudinal section, cross section of the strip with added 0.18 wt% Al-Ti-B refiner and applied ultrasonic treatment; (**g–i**) the top surface, longitudinal section, cross section of the strip with added 0.12 wt% Al-Ti-B refiner and applied ultrasonic treatment; (**j–l**) the top surface, longitudinal section, cross section of the strip with added 0.09 wt% Al-Ti-B refiner and applied ultrasonic treatment.

It can be seen from the above comparison that under the same 0.18 wt% Al-Ti-B refiner addition, the average grain size of the ultrasonic cast rolling strip was significantly smaller than that of the conventional cast rolling strip, indicating that the ultrasonic bending vibration could effectively refine the grain microstructure of the cast rolling strips. When the Al-Ti-B refiner addition was reduced to 0.12 wt%, the average grain size of the ultrasonic cast rolling strip was still smaller than that of the conventional cast rolling strip, but when the Al-Ti-B refiner addition was reduced to 0.09 wt%, the grain sizes of the ultrasonic cast rolling strip also became coarse and uneven, and the average grain size exceeded that of the conventional cast rolling strip with 0.18 wt% Al-Ti-B refiner.

The microstructure of the 1060 aluminum alloy cast rolling strip after the subsequent cold rolling is shown in Figure 4. Compared with the conventional cold rolling strip, the ultrasonic cast rolling strip with 0.18 wt% Al-Ti-B refiner or 0.12 wt% Al-Ti-B refiner had an obvious fiber structure that was finer and more uniform, without obvious blocks; the 0.18 wt% Al-Ti-B refiner-added ultrasonic cast rolling strip had the finest fiber structure after cold rolling. However, the 0.09 wt% Al-Ti-B refiner-added ultrasonic cast rolling strip had the tendency to coarsen the fiber grains after cold rolling.

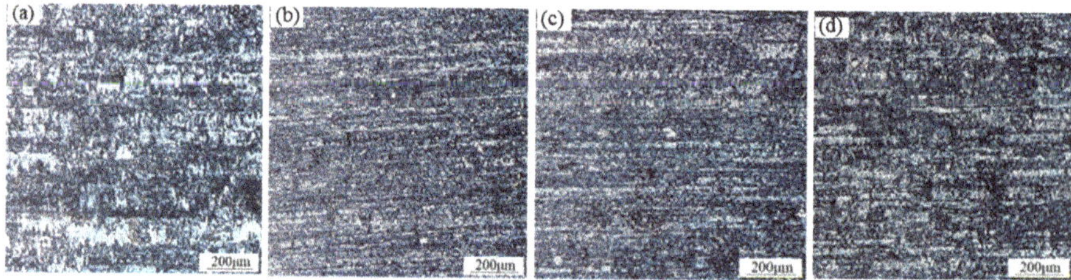

Figure 4. Grain structure after the subsequent cold rolling of the 1060 aluminum alloy cast rolling strip: (**a**) the top surface of the strip with added 0.18 wt% Al-Ti-B refiner and applied conventional treatment; (**b**) the top surface of the strip with added 0.18 wt% Al-Ti-B refiner and applied ultrasonic treatment; (**c**) the top surface of the strip with added 0.12 wt% Al-Ti-B refiner and applied ultrasonic treatment; (**d**) the top surface of the strip with added 0.09 wt% Al-Ti-B refiner and applied ultrasonic treatment.

3.2. Effect of the Ultrasonic Bending Vibration on the Precipitated Phases of the 1060 Aluminum Alloy Cast Rolling Strip and Subsequent Cold Rolling Strip

The precipitated phases on the surface, longitudinal section, and cross section of the 1060 aluminum alloy cast rolling strips under different treatments are shown in Figure 5. It can be seen that on the upper surface, longitudinal section, and cross section, the precipitated phase of the conventional cast rolling strip with 0.18 wt% Al-Ti-B refiner is coarse, unevenly distributed, and enriched, and the slag inclusion defects are very obvious; meanwhile, the precipitated phases in the ultrasonic cast rolling strip with 0.18 wt% Al-Ti-B refiner or 0.12 wt% Al-Ti-B refiner are more dispersed and uniformly distributed, and there are no large slag inclusion defects. However, when the addition amount of the Al-Ti-B refiner is reduced to 0.09 wt%, the precipitated phases of the ultrasonic cast rolling strip tend to be coarsened and the slag inclusion defects increase. This shows that the ultrasonic bending vibration can make the precipitated phases of the 1060 cast rolling strip disperse and uniformly distributed to a certain extent, and can significantly reduce the slag inclusion defects in the cast rolling strip.

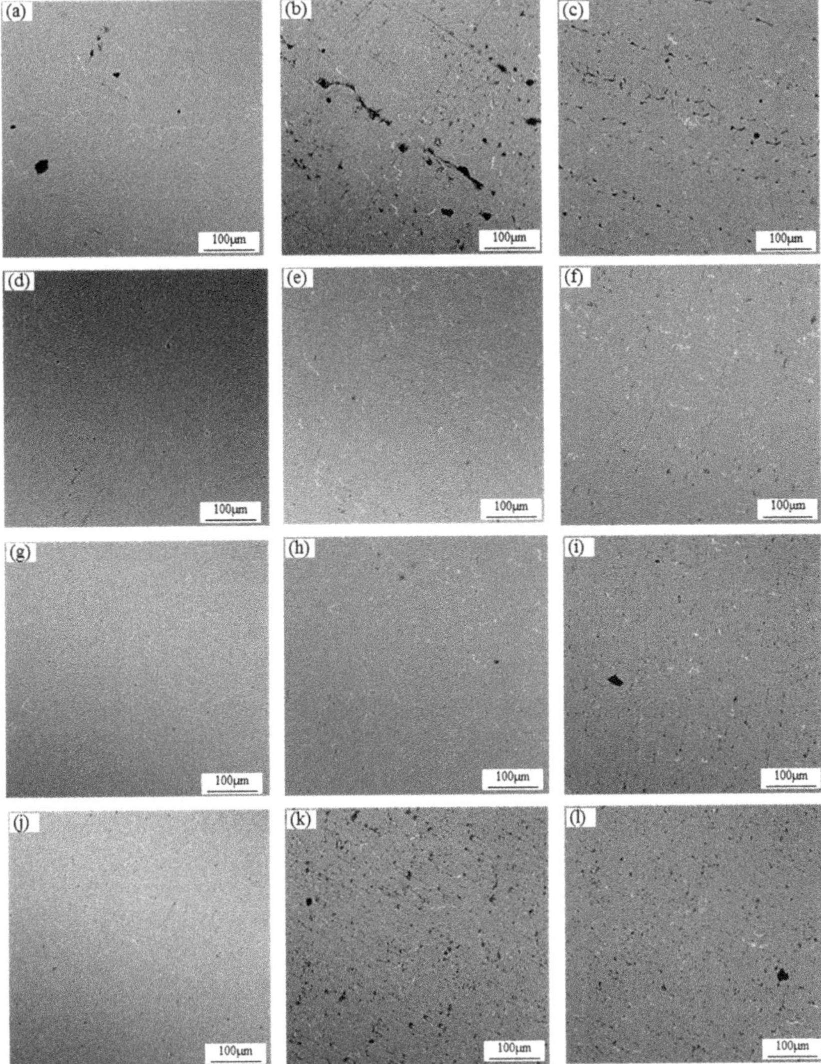

Figure 5. Precipitated phases of the 1060 aluminum alloy cast rolling strips: (**a–c**) the top surface, longitudinal section, cross section of the strip with added 0.18 wt% Al-Ti-B refiner and applied conventional treatment; (**d–f**) the top surface, longitudinal section, cross section of the strip with added 0.18 wt% Al-Ti-B refiner and applied ultrasonic treatment; (**g–i**) the top surface, longitudinal section, cross section of the strip with added 0.12 wt% Al-Ti-B refiner and applied ultrasonic treatment; (**j–l**) the top surface, longitudinal section, cross section of the strip with added 0.09 wt% Al-Ti-B refiner and applied ultrasonic treatment.

The precipitated phases on the surfaces of the 1060 aluminum alloy cast rolling strips after subsequent cold rolling under different treatments are shown in Figure 6. Compared with the subsequent cold rolling of the strip with conventional treatment, the ultrasonic cast rolling strip after cold rolling has a fine dispersed phase, a regular shape, and a relatively uniform distribution. This indicates that the solute elements in the material have uniformly diffused into the aluminum matrix, which can reduce the degree of component segregation. However, the size, shape, and distribution of the precipitated phases are not uniform, and there is an obvious solute enrichment area in the conventional cast rolling strip with the 0.18 wt% Al-Ti-B refiner addition after cold rolling. In addition, compared with the three types of ultrasonic cast rolling strips with the addition of the Al-Ti-B refiner, the ultrasonic cast rolling strip with added 0.18 wt% Al-Ti-B refiner or 0.12 wt% Al-Ti-B refiner has fewer slag inclusion defects after the cold rolling.

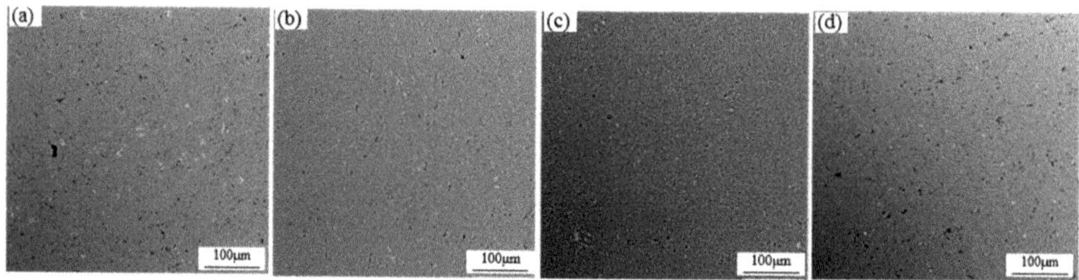

Figure 6. The precipitated phases of the subsequent cold rolling of the 1060 aluminum alloy cast rolling strip: (**a**) the top surface of the strip with added 0.18 wt% Al-Ti-B refiner and applied conventional treatment; (**b**) the top surface of the strip with added 0.18 wt% Al-Ti-B refiner and applied ultrasonic treatment; (**c**) the top surface of the strip with added 0.12 wt% Al-Ti-B refiner and applied ultrasonic treatment; (**d**) the top surface of the strip with added 0.09 wt% Al-Ti-B refiner and applied ultrasonic treatment.

3.3. The Effect of the Ultrasonic Bending Vibration on the Mechanical Properties of the 1060 Aluminum Alloy Cast Rolling Strips and Subsequent Cold Rolling Strips

The test results of the mechanical properties of the 1060 aluminum alloy cast rolling strips under four treatment conditions are shown in Table 2. It can be seen that in terms of tensile strength, adding the 0.18 wt% Al-Ti-B to the ultrasonic cast rolling strip improves the tensile property by 2%, while the other three have no significant difference. In terms of yield strength, compared with the addition of the 0.18 wt% Al-Ti-B refiner to the conventional cast rolling strip, the addition of the 0.18 wt% Al-Ti-B refiner, 0.12 wt% Al-Ti-B refiner, and 0.09 wt% Al-Ti-B refiner to the ultrasonic cast rolling strip caused a decrease of 0.98%, 0.98%, and 1.83%, respectively. In terms of elongation, compared to adding the 0.18 wt% Al-Ti-B refiner to the conventional cast rolling strip, adding the 0.18 wt% Al-Ti-B refiner, 0.12 wt% Al-Ti-B refiner, and 0.09 wt% Al-Ti-B refiner to the ultrasonic cast rolling strip caused an improvement of 6.42%, 2.73%, and 2.71%, respectively. Similarly, percentage reduction of area also increased by 2%, 2.52% and 1.57%, respectively. In summary, under the action of the ultrasonic bending vibration, the mechanical properties of the cast rolling strips have been improved to a certain extent. When the addition amount of the Al-Ti-B refiner was 0.18 wt%, the mechanical properties of the cast rolling strip had the greatest improvement.

The mechanical properties of the 1060 aluminum alloy cast rolling strip after cold rolling under four treatment conditions are shown in Table 3. It can be seen that in terms of tensile strength, compared with the conventional cast rolling strip with added 0.18 wt% Al-Ti-B refiner after cold rolling, the addition of the 0.18 wt% Al-Ti-B refiner, 0.12 wt% Al-Ti-B refiner, and 0.09 wt% Al-Ti-B refiner to the ultrasonic cast rolling strip after cold rolling all caused an improvement of 6.36%, 5.76%, and 1.68%, respectively. In terms of elongation, compared with the addition of the 0.18 wt% Al-Ti-B refiner to the conventional cast rolling strip after cold rolling, the addition of 0.18 wt% Al-Ti-B refiner to the ultrasonic cast rolling strip after cold rolling caused an increase of 4.11%, but the addition of the 0.12 wt% Al-Ti-B refiner or 0.09 wt% Al-Ti-B refiner to the ultrasonic cast rolling strip after cold rolling caused a reduction of 8.9% and 4.79%, respectively. This shows that the effect of the ultrasonic bending vibration on the mechanical properties of the cast rolling strip has been extended after the cold rolling, which is beneficial to the further processing of the cold rolling strip.

Table 2. Mechanical properties of the 1060 aluminum alloy cast rolling strips with different treatments.

Samples	Direction	Tensile Strength/MPa	Yield Strength/MPa	Elongation/%	Reduction in Area/%
0.18 wt% Al-Ti-B conventional	RD (cast rolling direction)	91.26	64.15	49.8	85.79
	45°	84.66	62.46	46.16	88.12
	TD (transverse cast rolling direction)	90.54	65.57	49.24	85.69
0.18 wt% Al-Ti-B ultrasound	RD	93.46	64.18	51.28	87.98
	45°	86.02	65.05	46.48	91.58
	TD	92.25	61.06	56.76	85.18
0.12 wt% Al-Ti-B ultrasound	RD	91.09	62.37	49.88	90.14
	45°	84.96	62.29	45.72	90.63
	TD	90.95	65.63	53.56	85.36
0.09 wt% Al-Ti-B ultrasound	RD	91.74	62.05	49.00	88.02
	45°	84.95	62.88	47.48	90.13
	TD	90.62	63.75	52.64	85.52

Table 3. Mechanical properties of subsequent cold rolling of the strips of the 1060 aluminum alloy cast rolling strip with different treatments.

Samples	Direction	Tensile Strength/MPa	Yield Strength/MPa	Elongation/%
0.18 wt% Al-Ti-B conventional	RD	150.73	148.10	2.92
	45°	145.06	141.40	3.8
	TD	153.51	153.00	2.04
0.18 wt% Al-Ti-B ultrasound	RD	167.02	165.48	2.96
	45°	153.5	149.58	3.08
	TD	157.38	156.61	3.08
0.12 wt% Al-Ti-B ultrasound	RD	165.4	165.15	2.61
	45°	156.75	154.94	3.25
	TD	153.02	153.02	2.12
0.09 wt% Al-Ti-B ultrasound	RD	152.85	152.22	1.85
	45°	149.7	145.92	2.52
	TD	154.3	153.53	3.96

3.4. Discussion

According to the experimental research in this paper, it can be concluded that the ultrasonic bending vibration improves the refinement efficiency of the Al-Ti-B refiner. Compared with the conventional casting and rolling strip with the 0.18 wt% Al-Ti-B refiner, the grain size of the ultrasonic casting and rolling strip with the 0.18 wt% Al-Ti-B refiner or 0.12 wt% Al-Ti-B refiner are smaller, the precipitated phases are dispersed, the distribution is uniform, and the mechanical properties are also improved. Moreover, after the cold rolling of the casting and rolling strip, the effect of the ultrasonic bending vibration on the improvement of the microstructure and properties of the aluminum alloy strip is inherited.

Al-Ti-B refiner plays an important role in the grain refinement of the aluminum alloy, and the refinement effect of the refiner is largely determined by the size and shape of the TiB_2 particles and $TiAl_3$ phase. The fine and uniformly distributed $TiAl_3$ phase and the dispersed and isolated TiB_2 particles are all helpful in improving the refining effect of the refiner [29,30]. TiB_2 particles inoculated into the melt through the Al-Ti-B master alloys are able to enhance the heterogeneous nucleation of α-Al grains [31–33]. However, TiB_2 particles are small in size and easy to aggregate into clusters, which are difficult to separate when added into the melt metal. Due to their high density and being easy to deposit, TiB_2 particles basically lose the role of the nucleation core, reducing the refining efficiency and becoming inclusions, thus reducing the quality of the aluminum products [34,35].

The ultrasonic bending vibration induced by the L-shaped ultrasonic rod into the aluminum melt in the launder will produce a significant acoustic flow effect, which will produce an obvious stirring and scouring effect in the aluminum melt. This causes the TiB_2 particles to be evenly distributed in the aluminum melt, reduces the aggregation and precipitation of the TiB_2, and makes the effect of the grain refinement more obvious. At the same time, the stirring effect of the ultrasonic bending vibration can promote the diffusion of the alloy elements in the aluminum melt and solid solution in the aluminum matrix, reduce the micro segregation, and inhibit the formation of coarse compounds, thus strengthening the matrix.

However, when the amount of Al-Ti-B refiner is reduced from 0.18 wt% to 0.09 wt%, even if it has the effect of the ultrasonic bending vibration the microstructure and properties of the 1060 aluminum alloy cast rolling strip and its subsequent cold rolling strip will deteriorate. This shows that the ultrasonic energy applied in the continuous casting and rolling process of the 1060 aluminum alloy in this paper cannot completely replace the role of the Al-Ti-B refiner. However, it can also be found from the text that when the amount of Al-Ti-B refiner is reduced from the conventional 0.18 wt% to 0.12 wt%, the microstructure and properties of the 1060 aluminum alloy cast rolling strip and its subsequent cold rolling strip are better than those of the conventional. The ultrasonic energy applied in this paper can reduce by the 1/3 amount of Al-Ti-B refiner used during continuous casting and rolling of the 1060 aluminum alloy, which is conducive to reducing the production costs.

4. Conclusions

(1) When the ultrasonic bending vibration is applied to the 1060 aluminum alloy during continuous casting and rolling under the condition of the same amount of Al-Ti-B refiner, the grains of the cast rolling strip are refined, the precipitated phases are dispersed and evenly distributed, the defects are obviously reduced, and the mechanical properties of the cast rolling strip are improved. The microstructure and properties of the ultrasonic cast rolling strip with 0.18 wt% Al-Ti-B refiner or 0.12 wt% Al-Ti-B refiner are better than those of the conventional cast rolling strip, but the microstructure and properties of the ultrasonic cast rolling strip with the 0.09 wt% Al-Ti-B refiner are worse than those of the conventional cast rolling strip.

(2) The microstructure and properties of the 1060 aluminum alloy ultrasonic cast rolling strips after cold rolling with 0.18 wt% Al-Ti-B refiner or 0.12 wt% Al-Ti-B refiner are better than those of the conventional cast rolling strip after cold rolling, but the microstructure and properties of the ultrasonic cast rolling strip with added 0.09 wt% Al-Ti-B refiner are worse than those of conventional cast rolling strip. This indicates that the effect of the ultrasonic bending vibration on the improvement of the microstructure and properties of the aluminum alloy strips is inherited during the cold rolling process, which is conducive to further processing.

(3) Applying the ultrasonic bending vibration during the continuous casting and rolling of the 1060 aluminum alloy can reduce the addition of Al-Ti-B refiner by 1/3. The L-shaped ultrasonic rod is resistant to high temperatures, and the transducer is not affected by the direct high temperature radiation of the aluminum melt. In addition, the nano ceramic radiator will not be eroded by long-term contact with the aluminum melt, which can realize continuous industrial production and reduce the production costs of industrial production.

Author Contributions: Conceptualization, C.S. and G.F.; validation, C.S., G.F. and X.M.; formal analysis, C.S. and G.F.; and writing—original draft preparation, G.F.; writing—review and editing, C.S.; funding acquisition, D.M. All authors have read and agreed to the published version of the manuscript.

Funding: The authors would like to acknowledge the financial assistance provided by Project of State Key Laboratory of High Performance Complex Manufacturing, Central South University (No.ZZYJKT2019-09), and National Basic Research Program of China (No.2014CB046702).

Conflicts of Interest: The authors declare no conflict of interest.

References

1. Cui, J.J.; Chen, L.Q.; L, Y.F.; Liu, J.H.; Yang, Y.F.; Xie, J.J. Effect of annealing on mechanical properties of low temperature cold rolling 1060 aluminum alloy. *Heat. Treat. Met.* **2018**, *11*, 111–116.
2. Sun, F.L.; Li, X.G.; Lu, L.; Cheng, X.Q.; Dong, C.F.; Gao, J. Corrosion behavior of 5052 and 6061 aluminum alloys in the deep sea environment of South China sea. *Acta Metall. Sin.* **2013**, *49*, 1219–1226. [CrossRef]
3. Pongen, R.; Birru, A.K.; Parthiban, P. Study of microstructure and mechanical properties of A713 aluminium alloy having an addition of grain refiners Al-3.5 Ti-1.5C and Al-3Cobalt. *Results Phys.* **2019**, *13*, 102105. [CrossRef]
4. Shivank, A.T.; Neha, B.; Prashant, V.; Smriti, M. Improving surface hardness of aluminum 6063 alloy using hardfacing. *Mater. Today* **2020**. [CrossRef]
5. Limmaneevichitr, C.; Eidhed, W. Novel technique for grain refinement in aluminum casting by Al–Ti–B powder injection. *Mat. Sci. Eng. R.* **2003**, *355*, 174–179. [CrossRef]
6. Zhang, J.X.; Zhong, J.H. Research status and analysis of master alloys for grain refinement of aluminum alloys. *Alum. Fabr.* **2002**, *1*, 24–26.
7. Lv, Y.N.; D, R.; Xu, G.M. Effect of refiner on microstructure and properties of 5052 aluminum alloy rolling. *Strip. Foundry. Technol.* **2018**, *6*, 1218–1220.
8. Liu, H.C. *Study on Al-Ti-B-Re Master Alloy Refiner*; Dalian University of Technology: Dalian, China, 2007.
9. Lan, H.F.; Guo, P.; Zhang, J.J. Effect of rare earth on refinement properties of Al-Ti-B-RE master alloy. *Foundry. Technol.* **2005**, *9*, 774–775+778.
10. Venkateswarlu, K.; Murty, B.; Chakraborty, M. Effect of hot rolling and heat treatment of Al–5Ti–1B master alloy on the grain refining efficiency of aluminium. *Mater. Sci. Eng. A* **2001**, *301*, 180–186. [CrossRef]
11. Doheim, M.A.; Omran, A.M.; Abdel-Gwad, A.; Sayed, G.A. Evaluation of Al-Ti-C Master Alloys as Grain Refiner for Aluminum and Its Alloys. *Met. Mater. Trans. A* **2011**, *42*, 2862–2867. [CrossRef]
12. Easton, M.; Qian, M.; Prasad, A.; StJohn, D. Recent advances in grain refinement of light metals and alloys. *Curr. Opin. Solid State Mater. Sci.* **2016**, *20*, 13–24. [CrossRef]
13. Qian, M.; Ramirez, A.; Das, A.; StJohn, D. The effect of solute on ultrasonic grain refinement of magnesium alloys. *J. Cryst. Growth* **2010**, *312*, 2267–2272. [CrossRef]
14. Wang, G.; Dargusch, M.; Qian, M.; Eskin, D.; StJohn, D. The role of ultrasonic treatment in refining the as-cast grain structure during the solidification of an Al–2Cu alloy. *J. Cryst. Growth* **2014**, *408*, 119–124. [CrossRef]
15. Gong, Y.-Y.; Luo, J.; Jing, J.-X.; Xia, Z.-Q.; Zhai, Q. Structure refinement of pure aluminum by pulse magneto-oscillation. *Mater. Sci. Eng. A* **2008**, *497*, 147–152. [CrossRef]
16. Liang, N.; Liang, Z.; Zhai, Q.; Wang, G.; StJohn, D. Nucleation and grain formation of pure Al under Pulsed Magneto-Oscillation treatment. *Mater. Lett.* **2014**, *130*, 48–50. [CrossRef]
17. Xia, C.X. *Industrial Experiment and Mechanism Research of Ultrasonic Continuous Casting and Rolling of 3003 Aluminum Alloy*; Central South University: Changsha, China, 2011.
18. Shi, C.; Shen, K. Twin-roll casting 8011 aluminium alloy strips under ultrasonic energy field. *Int. J. Light. Mater. Manuf.* **2018**, *1*, 108–114. [CrossRef]
19. Xu, G.M.; Pan, J.S. Effect of electromagnetic field on microstructure of 1100 aluminum alloy cast-rolling strip. *J. Northeast. Univ.* **2014**, *6*, 800–803.
20. Wang, C.N.; Connolley, T.; Tzanakis, L.; Eskin, D.; Mi, J.W. Characterization of Ultrasonic Bubble Clouds in A Liquid Metal by Synchrotron X-ray High Speed Imaging and Statistical Analysis. *Materials* **2019**, *13*, 44. [CrossRef]
21. Eskin, D.; Al-Helal, K.; Tzanakis, I. Application of a plate sonotrode to ultrasonic degassing of aluminum melt: Acoustic measurements and feasibility study. *J. Mater. Process. Technol.* **2015**, *222*, 148–154. [CrossRef]
22. Yang, Y.L.; Liu, Z.L.; Jiang, R.P.; Li, R.Q.; Li, X.Q. Microstructural evolution and mechanical properties of the AA2219/TiC nanocomposite manufactured by ultrasonic solidification. *J. Alloy. Compd.* **2019**, *11*, 151991. [CrossRef]
23. Puga, H.; Barbosa, J.; Machado, J.; Vilarinho, C. Ultrasonic grain refinement of die cast copper alloys. *J. Mater. Process. Technol.* **2019**, *263*, 336–342. [CrossRef]
24. Huang, J.; Li, J.; Li, C.; Huang, C.; Friedrich, B. Elimination of edge cracks and centerline segregation of twin-roll cast aluminum strip by ultrasonic melt treatment. *J. Mater. Res. Technol.* **2020**. [CrossRef]

25. Tian, Y.; Liu, Z.L.; Li, X.Q.; Li, R.Q.; Zhang, L.H.; Jiang, R.P.; Dong, F. The cavitation erosion of ultrasonic sonotrode during large-scalemetallic casting: Experiment and simulation. *Ultrason. Sonohem.* **2018**, *37*, 29–37. [CrossRef] [PubMed]

26. Dong, F.; Li, X.Q.; Zhang, L.H.; Ma, L.Y.; Li, R.Q. Cavitation erosion mechanism of titanium alloy radiation rods in aluminum melt. *Ultrason. Sonohem.* **2016**, *31*, 150–156. [CrossRef]

27. Ėskin, G.I.; Eskin, D.G. *Ultrasonic Treatment of Light Alloy Melts*; CRC Press/Taylor & Francis Group: Boca Raton, FL, USA, 2015.

28. Shi, C.; Wu, Y.; Mao, D.; Fan, G. Effect of Ultrasonic Bending Vibration Introduced by the L-shaped Ultrasonic Rod on Solidification Structure and Segregation of Large 2A14 Ingots. *Materials* **2020**, *13*, 807. [CrossRef]

29. Zhang, L.; Fang, Y. Al-Ti-B Master Alloy Refining Pure Aluminum Treatment. Available online: cprs.patentstar.com.cn (accessed on 24 December 2008).

30. Pattnaik, A.B.; Das, S.; Jha, B.B.; Prasanth, N. Effect of Al–5Ti–1B grain refiner on the microstructure, mechanical properties and acoustic emission characteristics of Al5052 aluminium alloy. *J. Mater. Res. Technol.* **2015**, *4*, 171–179. [CrossRef]

31. Ding, W.W.; Zhao, X.Y.; Chen, T.L.; Zhang, H.X.; Liu, X.X.; Cheng, Y.; Lei, D.K. Effect of rare earth Y and Al–Ti–B master alloy on the microstructure and mechanical properties of 6063 aluminum alloy. *J. Alloy. Compd.* **2020**, *830*, 154685. [CrossRef]

32. Ma, T.; Chen, Z.; Nie, Z.; Huang, H. Microstructure of Al-Ti-B-Er refiner and its grain refining performance. *J. Rare Earths* **2013**, *31*, 622–627. [CrossRef]

33. Nie, J.; Ma, X.; Ding, H.; Liu, X. Microstructure and grain refining performance of a new Al–Ti–C–B master alloy. *J. Alloy. Compd.* **2009**, *486*, 185–190. [CrossRef]

34. Birol, Y. The performance of Al–Ti–C grain refiners in twin-roll casting of aluminium foilstock. *J. Alloy. Compd.* **2007**, *430*, 179–187. [CrossRef]

35. Limmaneevichitr, C.; Eidhed, W. Fading mechanism of grain refinement of aluminum–silicon alloy with Al–Ti–B grain refiners. *Mater. Sci. Eng. A* **2003**, *349*, 197–206. [CrossRef]

Article

Effect of Ultrasonic Bending Vibration Introduced by the L-shaped Ultrasonic Rod on Solidification Structure and Segregation of Large 2A14 Ingots

Chen Shi [1,2,3,*], Yongjun Wu [1], Daheng Mao [3] and Gaofeng Fan [1]

1   College of Mechanical and Electrical Engineering, Central South University, Changsha 410083, China; llswyj@csu.edu.cn (Y.W.); fangaofeng@csu.edu.cn (G.F.)
2   State Key Laboratory of High Performance Complex Manufacturing, Central South University, Changsha 410083, China
3   Light Alloy Research Institute, Central South University, Changsha 410083, China; mdh@csu.edu.cn
*   Correspondence: shichen@csu.edu.cn; Tel.: +86-731-8887-7244

Received: 20 December 2019; Accepted: 5 February 2020; Published: 10 February 2020

Abstract: In order to achieve long-term and stable ultrasonic treatment in the direct chill semi-continuous casting process, a new L-shaped ceramic ultrasonic wave guide rod is designed to introduce ultrasonic bending vibration into 2A14 aluminum alloy melt. The effect of ultrasonic bending vibration on the solidification structure and composition segregation of large 2A14 aluminum alloy ingots (φ 830 mm × 6000 mm) in the process of semi-continuous casting were studied by means of a direct reading spectrometer, scanning electron microscope, metallographic microscope, and hardness test. The ultrasonic ingot treated by bending vibration was compared with the ingot without ultrasonic treatment and the ingot treated by the traditional straight-rod titanium alloy wave guide rod. The results show that, during the solidification of 2A14 aluminum alloy, ultrasonic treatment can significantly refine the grain, break up the agglomerated secondary phase, and make its distribution uniform. The macro-segregation degree of solute including the negative segregation at the edge of the ingots and the positive segregation in the center can be reduced. Through comparative analysis, the macrostructure of the ingot, treated by the L-shaped ceramic ultrasonic wave guide rod, was found to be better than that of the ingot treated by the traditional straight-rod titanium alloy wave guide rod. In particular, the grain refinement effect at the edge of the ingot was the best, the secondary phase was smaller, more solute elements can be dissolved into the α-Al matrix, and the ability of the L-shaped ultrasonic wave guide rod to restrain segregation was stronger at the edge of the ingot.

Keywords: L-shaped ultrasonic wave guide rod; ultrasonic bending vibration; 2A14 aluminum alloy; solidification structure; composition segregation

---

1. Introduction

A 2A14 aluminum alloy is a typical Al-Cu-Mg-Si alloy. Because of its high copper content, it has high strength and belongs to high strength duralumin [1–3]. It has good heat resistance, forging and welding properties, and it can be manufactured into free forgings, and die forgings with complex shapes, which are widely used in the field of aerospace [4–8]. Direct chill semi-continuous casting is one of the most effective technologies to produce large aluminum alloy ingots [9]. Due to the uneven distribution of temperature field and flow field in the solidification areas during casting, there are many problems in the large-scale aluminum alloy ingots, produced by traditional casting methods, which can easily cause an uneven distribution of solidification structure and solute composition, and defects, such as coarse grain, composition segregation, hot cracking, and shrinkage cavities. In particular, macro-segregation directly affects the quality of ingots and the subsequent finished products ratio of forging, and increases the material processing loss and production cost [10–13].

At present, applying appropriate ultrasonic vibration in the process of metal solidification is a good method to reducing casting defects, obtaining good structure and improving mechanical properties of materials [14–17]. Abramov V et al. [18] studied the effect of the ultrasonic treatment of the water-cooled transducer on the microstructure and properties of different industrial aluminum based alloy. The effect of ultrasonic treatment on the microstructure of as-cast alloy can be summarized as follows: Reduction of mean grain size, variation of phase distribution, and better material homogeneity and segregation control. Eskin et al. [19] introduced ultrasonics into the semi-continuous casting of aluminum alloy, produced AA2324 aluminum alloy round ingot with a diameter of 1200 mm. It was found that the grain after ultrasonic treatment was generally refined, and the solidification structure was non-dendrite. Li Xiaoqian et al. [20–23] studied the principle and mechanism of ultrasonic treatment with the straight-rod wave guide rod, analyzed the influence of ultrasonic power, frequency, and insertion depth of ultrasonic rods on the quality of large-diameter ingot, carried out a large number of casting experiments, and successfully obtained 2XXX and 7XXX ultrasonic large-diameter aluminum alloy ingots of various sizes.

However, there are some problems in the ultrasonic treatment process, which hinder its wide application in metallurgical industry. The traditional ultrasonic radiation rod, made of titanium alloy, was seriously corroded in the molten metal, thereby, reducing its service life and polluting the aluminum melt [24–26]. Moreover, the ultrasonic transducer, connected with the traditional straight-rod titanium alloy wave guide rod, is directly above the high-temperature metal melt, and the melt produces direct high-temperature heat radiation to the ultrasonic transducer. This often leads to the detuning problem of an ultrasonic vibration system or even damage of the transducer during the casting process, so it is difficult to achieve long-term stable ultrasonic treatment [27,28]. In this work, to overcome these problems, a new L-shaped ceramic ultrasonic wave guide rod was designed to introduce ultrasonic bending vibration into 2A14 aluminum alloy melt. The differences between treatment by the L-shaped ultrasonic wave guide rod and treatment by the traditional straight-rod titanium alloy wave guide rod were compared from the aspects of macrostructure, micro-grain size and morphology, secondary phase characteristics and composition segregation. The purpose of this paper is to verify the effectiveness of applying the L-shaped ultrasonic wave guide rod on the preparation of homogeneous large-scale aluminum alloy ingots from the experimental point of view, and then to find out the optimal ultrasonic introduction process, in order to provide certain theoretical guidance for industrial production.

2. Material and Methods

2.1. Experimental Equipment

A new type of ultrasonic guided wave device which mainly composed of (1) a magnetosteric transformer, (2) a L-shaped solid horn and (3) a tool head was designed, as shown in the Figure 1. The working frequency of ultrasonic was 19 ± 0.2 kHz, the power variation range was 0–1000 W, and the amplitude was 10 μm. The L-shaped horn was composed of the first level horn, which propagated the longitudinal vibration horizontally, and the second level horn, which propagated the bending vibration vertically. The first level horn and the second level horn, as well as the tool head, transducer and L-shaped horn were all connected by thread. The material of the horn was TC4 titanium alloy, and the material of the tool head in direct contact with the metal melt was nano-silicon nitride ceramics. The experimental device adopted the L-shaped ultrasonic wave guide rod, which made the ultrasonic transducer not directly above the high-temperature molten metal, and effectively avoided the direct radiation of high-temperature heat flow to the ultrasonic transducer, so as to improve the working stability and reliability of the ultrasonic vibration system. Moreover, the L-shaped solid horn transformed the sine wave transmitted longitudinally into the distorted wave transmitted transversely, so that the tool head inserted into the metal melt had strong ultrasonic transmission on its end face and side, thus strengthening the mechanical vibration and acoustic streaming of ultrasonic in the melt.

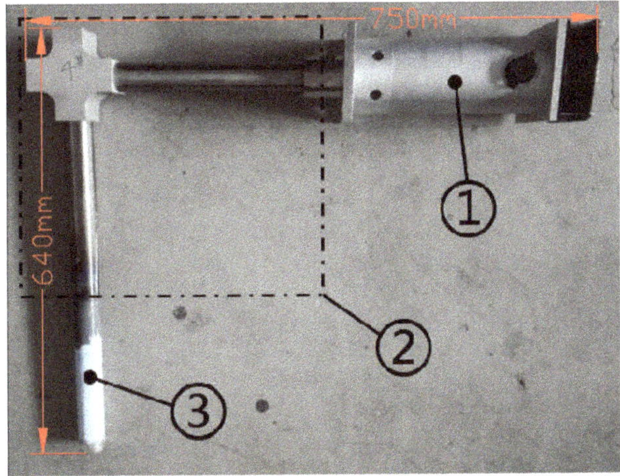

Figure 1. Structure of ultrasonic guided wave device.

2.2. Materials

2A14 aluminum alloy is an aluminum alloy with Cu, Mg, and Si as the main alloying elements. The experimental material was the 2A14 aluminum alloy with the chemical compositions listed in Table 1.

Table 1. Chemical compositions of the 2A14 aluminum alloy (wt.%).

Element	Si	Cu	Mg	Zn	Mn	Ti	Ni	Fe	Al
Content	0.6–1.2	3.9–4.8	0.4–0.8	≤0.30	0.4–1.0	≤0.15	≤0.10	0.0–0.7	Bal.

2.3. Experimental Process

The technological process of this ultrasonic casting experiment was as follows: The first step was to bake the smelting furnace, and then about 10 tons of commercial purity aluminum were put into the furnace. The material was heated and melted, and then fully stirred by electromagnetic stirrer. After the surface scum of the aluminum liquid was removed, metal additives, like Cu and Mg, as well as master alloy were added into the aluminum melt. Then, the sample was sent to the analysis room for composition detection and analysis. After the compositions of the molten aluminum alloy reached the required standard, Al-Ti-B wire rod was added. The molten aluminum was drained into the hot top mold through the launder, and the casting started after the hydrogen measurement of the melt was qualified. As shown in Figure 2a,b, ultrasonic wave was introduced into the aluminum melt from the top directly above the center to realize ultrasonic casting. Finally, the large ingot, as shown in Figure 2c, was transferred to the soaking pit furnace for homogenization treatment for 72 h. After annealing treatment, the chill layer of about 10 mm thick on the surface of ingots was removed by lathe. The specific casting parameters are shown in Table 2. In order to evaluate the effect of ultrasonic guided wave device ultrasonic treatment, the specific experimental conditions and variables were set as follows: (1) no ultrasonic treatment; (2) the pre-heated L-shaped ultrasonic wave guide rod was immersed 70 mm below the liquid level of the melt, and the ultrasonic treatment with a power of 500 W was applied; (3) the pre-heated the straight-rod wave guide rod was immersed 70 mm below the liquid level of the melt, and the ultrasonic treatment with a power of 500 W was applied.

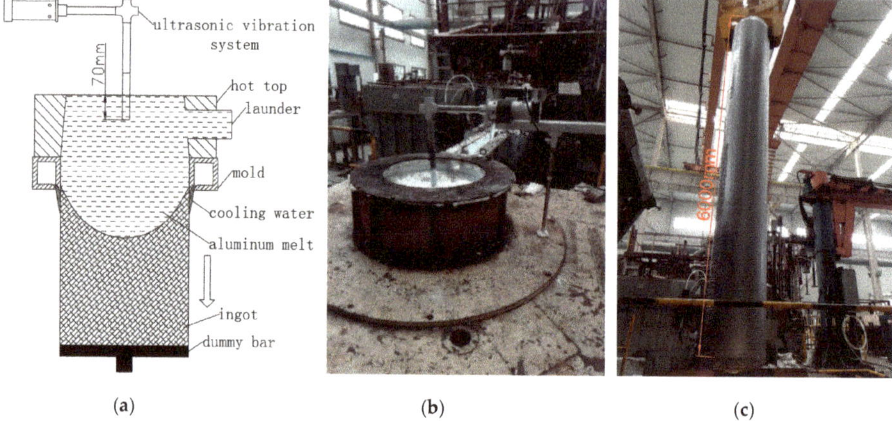

(**a**)   (**b**)   (**c**)

Figure 2. 2A14 aluminum alloy ultrasonic semi-continuous casting: (**a**) schematic diagram of casting; (**b**) photo of casting site; and (**c**) ingot product.

Table 2. Casting process parameters of the 2A14 aluminum ingots.

Size/mm	Pouring Temperature/°C	Cooling Temperature/°C	Speed of Water Flow/(L/min)	Speed of Introducing Ingot/(mm/min)
φ 830 × 6000	706	24	540	22

A circular section with a thickness of 20 mm was cut at 400 mm away from the head of ingots with a vertical band saw. A quarter of the section was taken for macrostructure analysis, and a long strip with a width of 100 mm was cut along the radius direction for spectral analysis. The sampling locations is shown in Figure 3. The specific analysis contents were as follows:

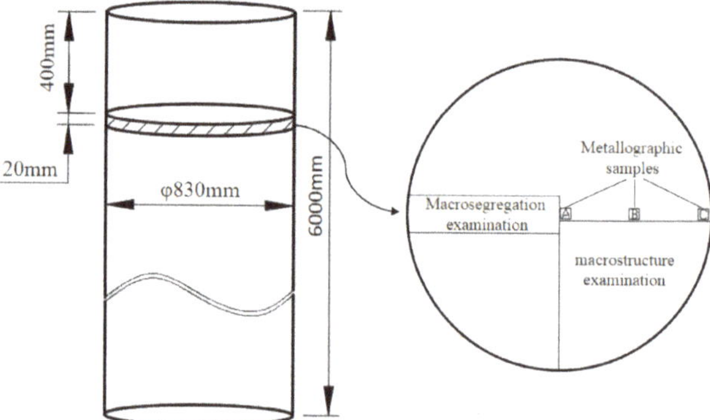

Figure 3. Sampling locations of ingot samples.

Firstly, the macrostructure of the samples was analyzed. After milling, the samples were polished with fine sandpaper, then the surface was cleaned with anhydrous ethanol, and the surface was treated with Keller's solution for 10–20 s to observe the corrosion effect. Then, the strip sample was milled, and the compositions were detected along the radius direction by the SPECTRO MAXx direct reading spectrometer. A group of data were measured every 40 mm. Each group of data measured three points and took the average value of them. Finally, the distribution curve of the alloy elements in the ingots along the radial direction was drawn by using the measured data. Moreover, the microstructure analysis was carried out. The sampling positions are shown in Figure 3. The positions A, B, and C in the figure correspond to the structure of the center, the one half radius, and the edge of the ingot samples, respectively. After the selected samples were polished, they were wiped with a cotton ball. The secondary phase structure and energy spectrum analysis were carried out using a Phenom fully automatic scanning electron microscope (Phenom-world BV, Holland). Then, their Vickers hardness values were measured using a HV-1000A microhardness tester. The applied load was 1 N with the dwell time of 10 s. Five results were tested for every samples, and after the maximum and minimum values were discarded, an average of the remaining three values was applied. Finally, the metallographic microstructure of the samples corroded by Keller's solution was observed using the OLYMPUS DSX500 metallurgical microscope (OLYMPUS Corporation, Japan), and the photos were taken, and the grain sizes were measured by the OLYMPUS Stream image analysis software. The three-circle intercept procedure was adopted. The measuring grid consists of three concentric circles of equal distance. Their total perimeter is 500 mm. This grid was used to measure at least five different fields of view of any choice for each specimen. The intercept points were counted manually with a mouse, and the average of the line intercept was calculated automatically by the analysis software.

3. Results and Discussions

3.1. Effect on Macrostructure

Figure 4 shows the macrostructure of the three casting ingots under different casting processes. Figure 4a shows that there are many dendrites, and coarse grains in the macrostructure of the ingot, without ultrasonic treatment. By comparing the images from the a-1 to the a-5 of the Figure, it was found that the grain near the edge of the ingot is the smallest, and the grain size from the edge to the center of the ingot is increasing. This is because the direct effect of cooling water, near the edge of the mold, made the under-cooling of the aluminum liquid in the edge area increase sharply, which induced homogeneous nucleation, thus, forming a large amount of crystal nuclei and fine equiaxed grains. However, in the center of the ingot, higher melt temperature produce coarser grains, mostly because of the de-activation of potent solidification sites. Comparing Figure 4a–c demonstrated that the macrostructure of the ingots with ultrasonic treatment is smaller and more uniform, compared with the ingot without ultrasonic treatment. Further comparison between Figure 4b and c shows that the effect of treatment by the L-shaped ultrasonic wave guide rod is slightly better than that of the traditional straight-rod wave guide rod on the whole. The grain size of the macrostructure of the ingots with ultrasonic treatment is smaller than that of the ingot without ultrasonic treatment, but the difference between them is not very obvious. This is mainly because a large amount of Al-Ti-B wire rod were added to the aluminum melt, as grain refiners during semi-continuous casting, which had achieved good grain refinement effect. Overall, the grain refinement effect under the synergistic effect of ultrasonic and grain refiners were significantly better than under by adding only grain refiners. The treatment by the L-shaped ultrasonic wave guide rod can not only significantly improve the refining ability of Al-Ti-B refiner, but also expand the effective range of the refiners.

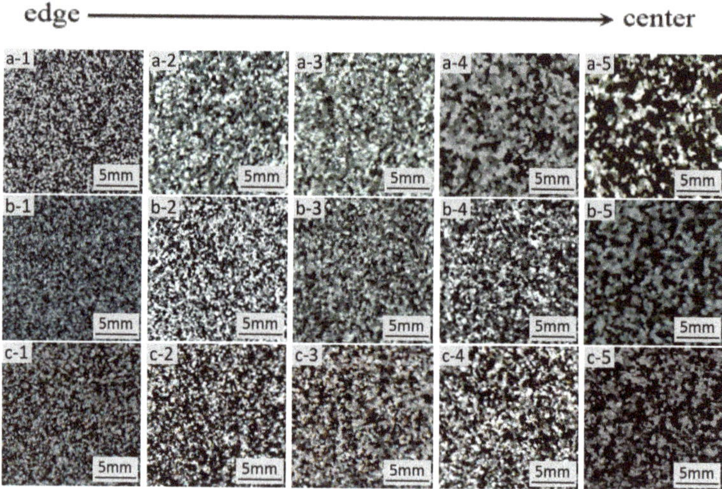

Figure 4. Macrostructure of obtained under different processing conditions: (**a**) no ultrasonic treatment; (**b**) treatment by the L-shaped ultrasonic wave guide rod; (**c**) treatment by the straight-rod ultrasonic wave guide rod.

3.2. Effect on Microstructure

Figure 5 shows the microstructure of the ingots cross-section along the radius under different processing conditions. It can be seen from the figure that the grain size of the three ingots is decreasing from the center to the edge. The dendrites in the center of the conventional ingot are dominant, the secondary dendrite arms are large, and only a few fine grains exist. At the one half radius mark, the number of coarse dendrites decreases and the grain size is smaller. The grain is basically fine equiaxed at the edge. After ultrasonic treatment, the grain of the ingots is refined to a certain extent, mainly fine equiaxed grain with more uniform distribution, especially in the center and at the 1/2 radius. Obviously, the closer the distance from the ultrasonic tool head, the better the grain refinement effect, which shows that the effect of grain refinement of ultrasonic treatment is closely related to its range of action. From the distribution curve of grain size of the ingots, under different processing conditions in Figure 6, it can be seen that in the conventional ingot, the grain size in the center is the thickest with an average grain size of 251.74 μm. Under the action of the L-shaped ultrasonic wave guide rod and the straight-rod ultrasonic wave guide rod, the average grain sizes decreased to 190.62, and 185.81 μm, respectively, and the decreasing amplitude exceeded 60 μm. At the edge, the average grain size of the ingot treated by the L-shaped ultrasonic wave guide rod is 135.96 μm, and the refining effect is better than that of the ingot treated by the straight-rod ultrasonic wave guide rod. The mechanism of ultrasonic grain refinement can be attributed to the cavitation effect and acoustic steam of ultrasonic [29]. The L-shaped ultrasonic wave guide rod had end radiation, but also more side radiation, compared with the action of the straight-rod ultrasonic wave guide rod, which made the area of ultrasonic cavitation larger. The shock wave generated in the process of cavitation bubbles collapse had a strong crushing effect on the primary dendrite and growing dendrite structure in the aluminum alloy melt, which increased the number of crystal nucleus during the solidification process, and the cavitation effect led to instantaneous local lower cold, in part, and heterogeneous nucleation. At the same time, the instantaneous high-pressure, produced by the collapse and fracture of the cavitation bubbles, constantly impacted the surface of heterogeneous particles in the melt, increased the contact angle with aluminum liquid, improved its wettability, activated the heterogeneous particles, and promoted the nucleation core to increase during solidification and crystallization. A large amount of nuclei was generated in these areas, thereby, increasing the number of nucleation and refining microstructure [23].

There was still a certain sound flow effect at the edge of the ingots, which made the aluminum melt be stirred continuously and forced the grain to disperse evenly in the melt. At the same time, it made the temperature field and solute field more uniform in the mold, and the growth direction of the grain became relatively uniform, so as to achieve the effect of refining grain.

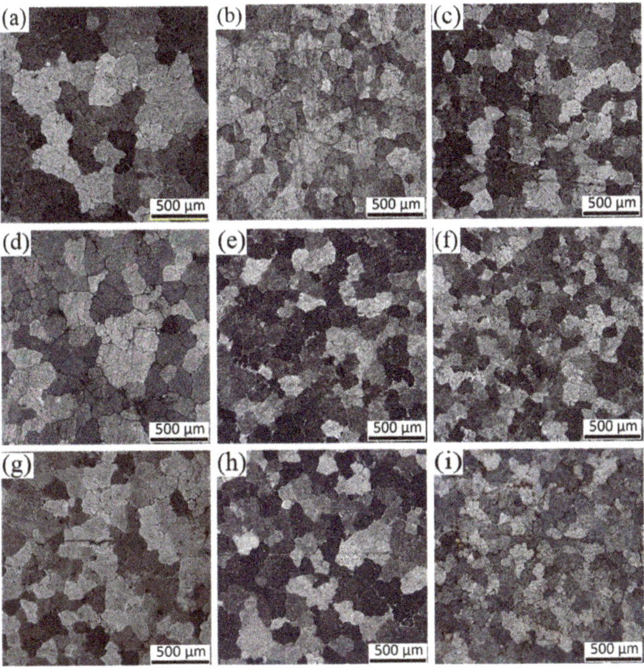

Figure 5. Microstructure of obtained under different processing conditions: (**a**) Center of ingot without ultrasonic; (**b**) 1/2 radius of ingot without ultrasonic; (**c**) edge of ingot without ultrasonic; (**d**) center of ingot by L-shaped ultrasonic; (**e**) 1/2 radius of ingot by L-shaped ultrasonic; (**f**) edge of ingot by L-shaped ultrasonic; (**g**) center of ingot by straight-rod ultrasonic; (**h**) 1/2 radius of ingot by straight-rod ultrasonic; and (**i**) edge of ingot by straight-rod ultrasonic.

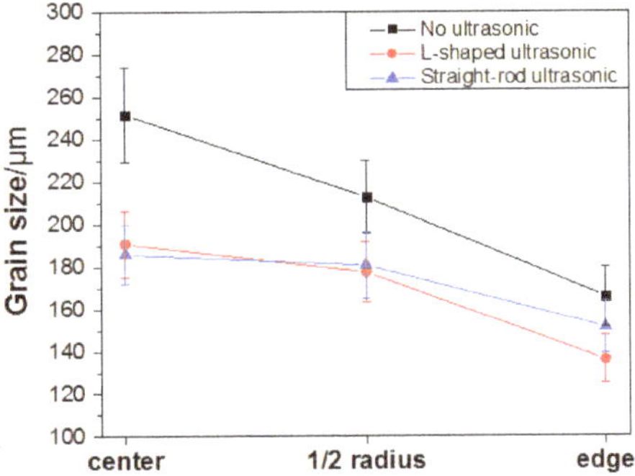

Figure 6. Distribution of grain size of ingots under different processing conditions.

3.3. Effect on the Secondary Phase

Figure 7 shows the distribution of the secondary phase along the radial direction of the ingots under different processing conditions. The dark gray area is α-Al matrix, and the white part is the secondary phase. In the solidification process of the alloy, the matrix phase formed first, and then the alloy elements precipitated, due to the decrease in solubility of the matrix, forming binary or ternary eutectic phase structure with Al element. It can be seen from the figure that the grain size in the microstructure of the conventional ingot is obviously larger, and there are some dendrites, while the proportion of equiaxed grain in the ingots by ultrasonic treatment is larger, the grain becomes smaller and more uniform, the size of the secondary phase is also reduced, the morphology is improved, and the distribution is more uniform. All of these significantly improved the comprehensive mechanical properties of the ingots. In the center of the ingots, whether it is the conventional ingot, without ultrasonic treatment, or the ingots with ultrasonic treatment, the secondary phase agglomeration phenomenon at the grain boundaries is very obvious and there is still serious segregation. At the one half radius of the ingots, the ultrasonic treatment broke-up some of the clustered secondary phase, and branched intergranular secondary phase with a discontinuous distribution appeared. In particular, the crushing effect of the ingot, treated with the L-shaped ultrasonic wave guide rod, was found to be better than that of ingots treated with the straight-rod ultrasonic wave guide rod, and the coarse secondary phase on the grain boundary was reduced, but there are still obvious long strip and massive eutectic phases. At the edge of the ingots, there are relatively few agglomerations of the secondary phase, and the secondary phase is mainly in the form of fine mesh. The secondary phase of the ingot, treated by the straight-rod wave guide rod, is increasingly fine, distributed in an intermittent network at the grain boundary of equiaxed grain, and has been basically dissolved in the matrix in some areas.

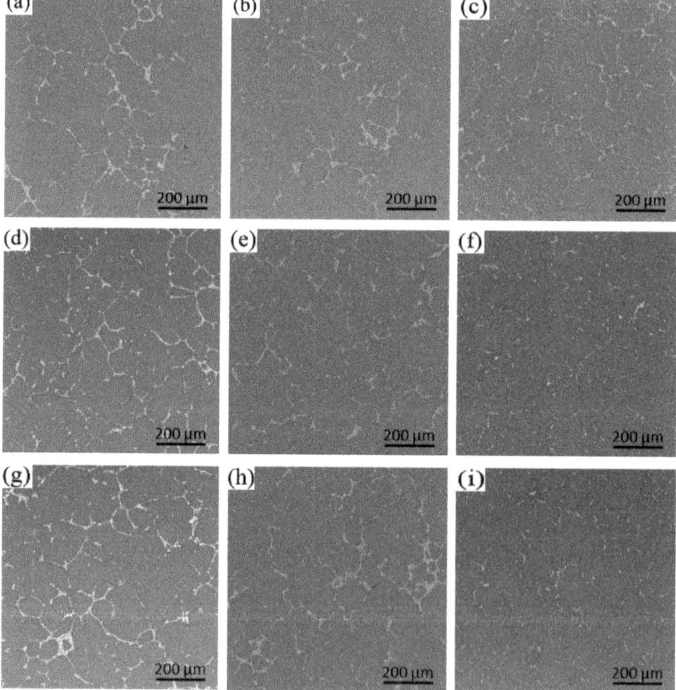

Figure 7. Distribution of the secondary phase of the ingots under different processing conditions: (**a**) Center of ingot without ultrasonic; (**b**) 1/2 radius of ingot without ultrasonic; (**c**) edge of ingot without ultrasonic; (**d**) center of ingot by L-shaped ultrasonic; (**e**) 1/2 radius of ingot by L-shaped ultrasonic; (**f**) edge of ingot by L-shaped ultrasonic; (**g**) center of ingot by straight-rod ultrasonic; (**h**) 1/2 radius of ingot by straight-rod ultrasonic; and (**i**) edge of ingot by straight-rod ultrasonic.

Figures 8 and 9 show the distribution and the area percentage of the secondary phase in the grain at the 1/2 radius respectively. The coarse bars and blocks of the secondary phase in the conventional ingot without ultrasonic treatment have larger proportion and are dispersed in the grain. The secondary phase in the grain of the ingots by ultrasonic treatment is mainly granular and acicular, and the distribution is relatively uniform, but there are still a great deal of little block or short rod-like secondary phase. The application of ultrasonic treatment can refine the secondary phase in the grain, break and re-melt the primary dendrite in the solidification front of the melt, reduce its aggregation and growth, and promote its uniform distribution.

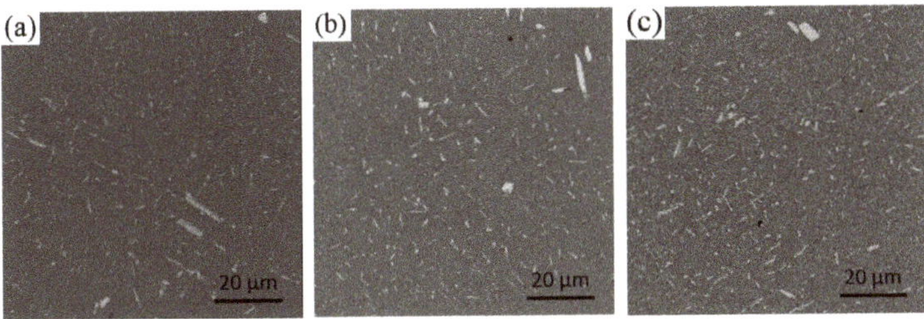

Figure 8. Distribution of the secondary phase in the grain at the 1/2 radius: (**a**) No ultrasonic treatment; (**b**) treatment by the L-shaped ultrasonic wave guide rod; (**c**) treatment by the straight-rod ultrasonic wave guide rod.

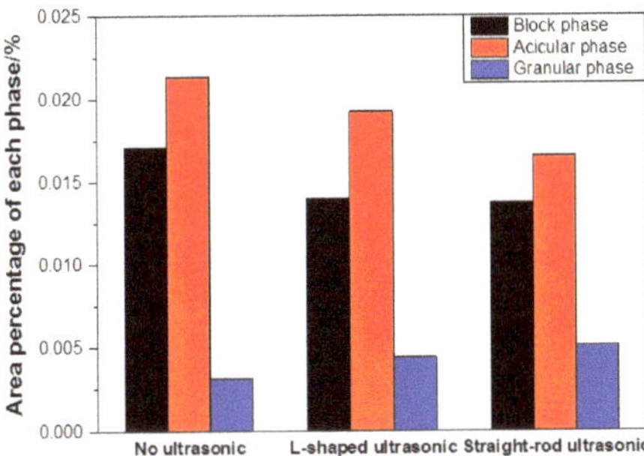

Figure 9. Area percentage of each secondary phase in the grain at the 1/2 radius.

Figure 10 shows the results of line scan analysis of main alloy elements in the ingots under different processing conditions. It can be seen from the figure that the main alloy elements Cu, Mg, and Si, in the ingots without ultrasonic treatment, have obvious segregation at the grain boundary, and the segregation of Cu element is particularly serious. Also, Cu, Mg, and Si are mainly enriched in the secondary phase, Al_2Cu, which is white. In the ingots treated by ultrasonic, the concentration of main alloy elements decreased, solute diffusion was promoted, part of Al_2Cu phase dissolved in the matrix, but a large amount number of Cu element were still segregated at the grain boundary.

Further point scanning analysis was carried out on the α-Al matrix structure of the ingots under different processing conditions. About 100 points in one grain were selected for scanning, and the average value of the measured 100 component points was taken as the main content of solute elements in the matrix. Table 3 shows the content of solute elements in the α-Al matrix structure of the ingots treated with different processing conditions.

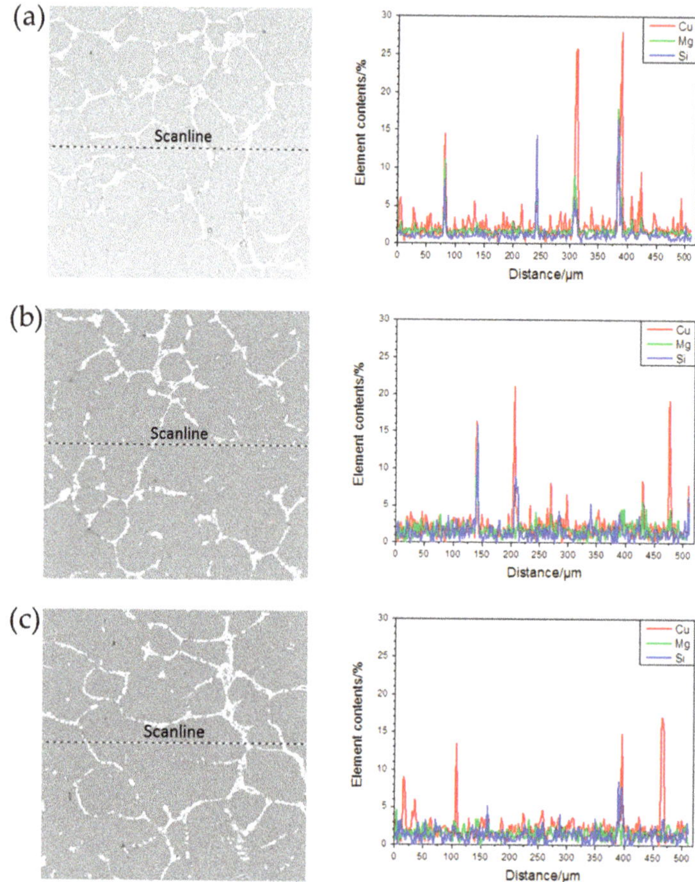

Figure 10. Results and areas of line scan analysis of main alloy elements in the ingots under different processing conditions: (**a**) No ultrasonic treatment; (**b**) treatment by the L-shaped ultrasonic wave guide rod; and (**c**) treatment by the straight-rod ultrasonic wave guide rod.

Table 3. Content of solute elements in the α-Al matrix structure of the ingots treated under different processing conditions (Wt.%).

Solute Elements	No Ultrasonic	L-Shaped Ultrasonic	Straight-Rod Ultrasonic
Cu	3.40	3.65	3.68
Si	0.89	1.21	1.17
Mg	0.94	1.07	1.09

3.4. Effect on MacroSegregation

The segregation rate ΔC was calculated by the following equation,

$$\Delta C = \frac{C_i - C_0}{C_0} \tag{1}$$

where C_i is the mean composition at a specific location, C_0 is the average (or nominal) alloy composition. ΔC > 0 represents positive segregation while ΔC < 0 represents negative segregation. The segregation index S is introduced to measure the overall macrosegregation degree of ingots in radial direction,

$$S = \Delta C_{max} - \Delta C_{min} \tag{2}$$

where ΔC_{max} and ΔC_{min} are the maximum, and minimum, segregation rates in the selected direction, respectively. The larger the S value is, the greater the macrosegregation degree of elements is, and the more uneven the solute distribution is.

Figure 11 shows the variation of segregation rate of Cu, Mg, and Si elements in the cross-section of the ingots under different conditions, which can clearly and intuitively reflect the macrosegregation of the ingots. The ingot without ultrasonic treatment has a large positive segregation in the center, and a very serious negative segregation near the edge, and the composition distribution curve has a sharp drop in the melt at the position. The segregation index of Cu, Mg, and Si decreases from 0.078, 0.053 and 0.072 to 0.048, 0.038 and 0.055, respectively after ultrasonic treatment of the L-shaped ultrasonic wave guide rod within 100 mm from the center. The segregation index of Cu, Mg, and Si decreases from 0.086, 0.069, and 0.036 to 0.044, 0.025, and 0.014, respectively after ultrasonic treatment of the straight-rod ultrasonic wave guide rod within 100 mm at the 1/2 radius. The segregation index of Cu, Mg, and Si changes from 0.212, 0.016 and 0.276 to 0.143, 0.120 and 0.132, respectively after ultrasonic treatment of the L-shaped ultrasonic wave guide rod near the edge. The application of ultrasonic treatment can effectively reduce the degree of solute segregation, reduce the negative segregation at the edge of ingots, reduce the positive segregation in the center, and make the distribution of solute elements in different parts of ingots tend to be uniform. However, the effect of the ultrasound is still limited, and the distribution of Cu and Si elements in the surface and center areas is still uneven, and there is a large gap between the two areas. The ability of the L-shaped ultrasonic wave guide rod in restraining segregation is stronger in the side of ingots, and the ability of the straight-rod ultrasonic wave guide rod to restrain segregation is stronger in the center and at the one-half radius of ingots. This is mainly because the L-shaped ultrasonic wave guide rod had not only the end radiation, but also more side radiation compared with the action of the straight-rod ultrasonic wave guide rod. Especially, micro jet with high velocity was formed on the side of the ultrasonic rod, which forced the local melt in the mold to produce convection in the radial direction and played a certain role in stirring the aluminum melt. Stirring reduced the solidification speeds at the edge of the ingots to a certain extent, so that the components in the solidification process had more time to diffuse, thus, greatly unifying the solute field, reducing the concentration gradient at the solidification front, as well as the concentration of solute elements. Under the action of strong shock wave and micro jet produced by cavitation effect, the high solute liquid at the solidification front was rapidly diffused, thus reducing the macro negative segregation at the edge of the ingots [30]. However, due to the rapid attenuation of ultrasound in the liquid, the effect of ultrasonic at the edge of the mold was very weak, compared with that of the center. Although, it can improve the inverse segregation of solute elements to a certain extent, it cannot be eliminated fundamentally, and the effect is limited. During the solidification process, the positive segregation on the surface of ingots occurred, while the negative segregation near the surface was caused by the deformation of dendrite network [31]. In the surface area of the ingots, cooling water was directly sprayed on it, the cooling rate was high, and the temperature gradient in solidified shell was very large. The thermal shrinkage and volume deformation of the dendrite network at the edge of ingots were larger than the internal area of ingots, and the resulting liquid-phase flow towards the ingot surface was also very obvious, while, the intergranular melt flowed continuously in the direction of shrinkage. When the ingot moved to the lower part of the mold, the solidification latent heat also made the rigid mushy zone on the ingot surface partially re-melt. The melt rich in solute elements between dendrites moved to the shell of the ingots along the gaps driven by extrusion pressure, which resulted in poor solute area near the surface of ingots and negative segregation. However, the solute elements were highly enriched on the surface of the ingots, and the composition was significantly higher than the average value, forming obvious segregation tumor, which reduced the quality of ingot surface.

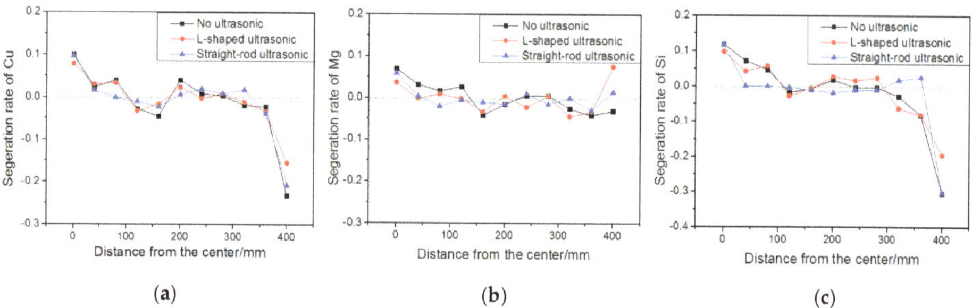

Figure 11. Variation of segregation rate of Cu (**a**), Mg (**b**), and Si (**c**) elements in the cross section of ingots under different processing conditions.

3.5. Effect on Mechanical Property

Hardness is a comprehensive index to measure the strength, plasticity, fracture toughness, deformation resistance, and other mechanical properties of aluminum alloy. Figure 12 shows the hardness test results, along the radial direction of the ingots, under different processing conditions. It can be seen from the figure that the hardness values of every position of the ingots, with ultrasonic treatment, is higher than that of the ingot without ultrasonic treatment, and the values tend to be consistent. The hardness values rose greatly in the center of ingots treated with the L-shaped ultrasonic wave guide rod and the straight-rod ultrasonic wave guide rod, increasing by 4.8%, and 5.6%, respectively. Previous studies have shown that, when ultrasonic treatment is applied to the solidification process of the aluminum alloy, the cavitation effect and acoustic steam increase the nucleation rate, refine the grain, and make the solidification structure more compact and uniform. At the same time, porosities, inclusions, and other defects are also effectively suppressed, and the volume fraction and size of the secondary phase is reduced, the morphology and distribution of the secondary phase are improve, and the composition segregation can also be restrained to a certain extent. The comprehensive effect of these factors effectively improved the mechanical properties of ultrasonic ingots. Compared with other positions, the effect of grain refinement in the center of ingots was better, the secondary phase was smaller and more evenly distributed, and the hardness of aluminum alloy was greatly improved. At the edge of the ingots, the fine equiaxed crystal structure increased the resistance of dislocation movement, thus, increasing the strength. However, due to the existence of serious negative segregation, the decrease of Cu content resulted in the decrease of the amount of Al_2Cu in the matrix, and the hardness values of aluminum alloy were not much higher than that at the 1/2 radius of the ingots. The mechanical properties and isotropy of the ingots, treated by ultrasonic, improved, which is beneficial to the subsequent processing.

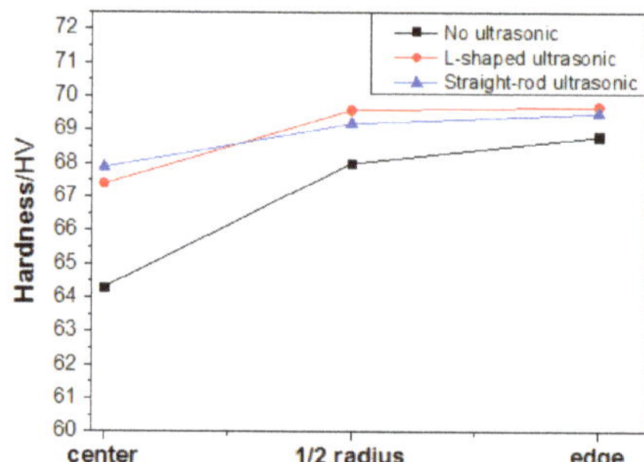

Figure 12. Hardness test results along the radial direction of the ingots under different processing conditions.

4. Conclusions

(1) The macrostructure of the ingots, with ultrasonic treatment, is smaller and more uniform. The L-shaped ultrasonic wave guide rod had not only the end radiation, but also more side radiation, compared with the action of the straight-rod ultrasonic wave guide rod and the effect of treatment, is better at the edge.

(2) Ultrasonic treatment can break the agglomerated secondary phase, and the secondary phase of the ingots treated by the L-shaped ultrasonic wave guide rod is smaller, and more solute elements are dissolved into the α-Al matrix.

(3) Ultrasonic treatment can effectively decrease the degree of solute segregation. The ability of the L-shaped ultrasonic wave guide rod to restrain segregation is stronger at the edge of ingots, and the ability of the straight-rod ultrasonic wave guide rod to restrain segregation is stronger in the center and at half a radius of ingots.

The effect is good in applying the L-shaped ultrasonic wave guide rod on the preparation of homogeneous large-scale aluminum alloy ingots, which will make it possible that the new L-shaped ceramic ultrasonic wave guide rod can be widely used in metallurgical industry. In our future research, the mechanical properties of the ingots, the resistance to high-temperature metal melt erosion and service life of the ultrasonic rod, and the working stability of the ultrasonic system, will be explored.

Author Contributions: Conceptualization, C.S. and Y.W.; validation, C.S., Y.W., and G.F.; formal analysis, C.S. and Y.W.; writing—original draft preparation, Y.W.; writing—review and editing, C.S.; funding acquisition, D.M. All authors have read and agreed to the published version of the manuscript.

Funding: This research received no external funding.

Acknowledgments: The authors would like to acknowledge the financial assistance provided by the Project of State Key Laboratory of High Performance Complex Manufacturing, Central South University (No. ZZYJKT2019-09), and National Basic Research Program of China (No. 2014CB046702).

Conflicts of Interest: The authors declare no conflict of interest.

References

1. Gavgali, M.; Aksakal, B. Effects of various homogenisation treatments on the hot workability of ingot aluminium alloy AA2014. *Mater. Sci. Eng.* **1998**, *254*, 189–199. [CrossRef]
2. Yuan, H.R.; Lin, S.B.; Yang, C.L.; Fan, C.L.; Wang, S. Microstructure and porosity analysis in ultrasonic assisted TIG welding of 2014 aluminum alloy. *Int. J. Lightweight Mater. Manuf.* **2011**, *20*, 39–43.
3. Sinan, A.; Bülent, B. Effects of Ageing and Cryo-ageing Treatments on Microstructure and Hardness Properties of AA2014–SiC MMCs. *Trans. Indian Inst. Met.* **2018**, *71*, 2035–2042.
4. Zhao, Y.H.; Lin, S.B.; He, Z.Q.; Wu, L. Microhardness prediction in friction stir welding of 2014 aluminium alloy. *Sci. Technol. Weld. Join.* **2006**, *11*, 178–182. [CrossRef]
5. Ghosh, K.S.; Tripati, K. Microstructural Characterization and Electrochemical Behavior of AA2014 Al-Cu-Mg-Si Alloy of Various Tempers. *J. Mater. Eng. Perform.* **2018**, *27*, 5926–5937. [CrossRef]
6. Ersin, A.G. An Investigation on the Effects of Pre-deformation and Temperature on Thixotropic Microstructure of AA2014 Aluminum Alloy. *Trans. Indian Inst. Met.* **2016**, *69*, 935–940.
7. Fu, J.; Jin, H.J.; Wu, S.J.; Zhang, J.Z. Effect of heat treatment on microstructure and properties of 2A14 aluminum alloy. *J. Trans. Mater. Heat. Treat.* **2016**, *37*, 189–194.
8. Goncalves, M.C.; Martins, M.G.; Misiolek, W.Z.; Van Geertruyden, W.H. Homogenization and Hot Workability of Alloy AA2014. *Mater. Sci. Forum* **2002**, *396–402*, 393–398. [CrossRef]
9. Luo, H.J.; Jie, W.Q.; Gao, Z.M.; Zheng, Y.J. Numerical simulation for macrosegregation in direct-chill casting of 2024 aluminum alloy with an extended continuum mixture model. *Trans. Nonferrous Met. Soc. China* **2018**, *28*, 1007–1015. [CrossRef]
10. Fezi, K.; Plotkowski, A.; Krane, M.J.M. Macrosegregation modeling during direct-chill casting of aluminum alloy 7050. *Numer. Heat Transf. Part A Appl.* **2016**, *70*, 1–25. [CrossRef]
11. Jamaly, N.; Haghdadi, N.; Phillion, A.B. Microstructure, macrosegregation, and thermal analysis of direct chill cast AA5182 aluminum alloy. *J. Mater. Eng. Perform.* **2015**, *24*, 2067–2073. [CrossRef]

12. Li, R.Q.; Liu, Z.L.; Chen, P.H.; Zhong, Z.T.; Li, X.Q. Investigation on the manufacture of a large-scale aluminum alloy ingot: Microstructure and macrosegregation. *Adv. Eng. Mater.* **2016**, *19*, 1600375. [CrossRef]

13. Peng, H.; Li, R.Q.; Li, X.Q.; Ding, S.; Chang, M.J.; Liao, L.Q.; Zhang, Y.; Chen, P.H. Effect of multi-source ultrasonic on segregation of Cu elements in large Al-Cu alloy cast ingot. *Materials* **2019**, *12*, 2828. [CrossRef] [PubMed]

14. Wang, G.; Dargusch, M.S.; Qian, M.; Eskin, D.G.; StJohn, D.H. The role of ultrasonic treatment in refining the as-cast grain structure during the solidification of an Al–2Cu alloy. *J. Cryst. Growth* **2014**, *408*, 119–124. [CrossRef]

15. Zhao, Z.F.; Qi, J.G.; Zhang, C.J.; Dai, S.; Wang, J.Z. Review on the Application of Liquid Metal Modification on the Study of Solidification Control. *J. Foundry Technol.* **2011**, *32*, 722–725.

16. Liang, G.; Shi, C.; Mao, D.H. Research Progress on External Energy Filed Removing Solidification Forming Defects of Steels. *J. Mater. Mech. Eng.* **2016**, *40*, 20–23.

17. Tzanakis, I.; Lebon, G.S.B.; Eskin, D.G.; Pericleous, K. Investigation of the factors influencing cavitation intensity during the ultrasonic treatment of molten aluminium. *J. Mater. Des.* **2016**, *90*, 979–983. [CrossRef]

18. Abramov, V.; Abramov, O.; Bulgakov, V.; Sommer, F. Solidification of aluminium alloys under ultrasonic irradiation using water-cooled resonator. *J. Mater. Lett.* **1998**, *37*, 27–34. [CrossRef]

19. Eskin, G.I.; Eskin, D.G. *Ultrasonic Treatment of Light Alloy Melts*, 2nd ed.; CRC Press: Boca Raton, FL, USA, 2014.

20. Liu, Z.L.; Li, R.Q.; Jiang, R.P.; Zhang, L.H.; Li, X.Q. Scalable ultrasound-assisted casting of ultra-large 2219 Al alloy ingots. *J. Metall. Mater. Trans. A* **2019**, *50*, 1146–1152. [CrossRef]

21. Jiang, R.P.; Li, X.Q.; Chen, P.H.; Li, R.Q.; Zhang, X. Effect and kinetic mechanism of ultrasonic vibration on solidification of 7050 aluminum alloy. *Aip Adv.* **2014**, *4*, 077125. [CrossRef]

22. Li, X.Q.; Li, R.Q.; Dong, F.; Chen, P.H.; Jiang, R.P. Simulation and experimental study of cavitation region caused by longitudinal and transverse vibration of casting ultrasonic radiator. *IOP Conf. Ser. Mater. Sci. Eng.* **2015**, *72*, 052052. [CrossRef]

23. Li, R.Q.; Li, X.Q.; Chen, P.H.; Guo, X.; Zhang, M. Effect rules and function mechanism of ultrasonic cavitation on solidification microstructure of large size high-strength aluminum alloy with hot top casting. *J. Cent. South Univ. (Sci. Technol.)* **2016**, *47*, 3360.

24. Tian, Y.; Liu, Z.L.; Li, X.Q.; Zhang, L.H.; Li, R.Q.; Jiang, R.P. The cavitation erosion of ultrasonic sonotrode during large-scale metallic casting: Experiment and simulation. *Ultrason. Sonochem.* **2018**, *43*, 29–37. [CrossRef] [PubMed]

25. Dong, F.; Li, X.Q.; Zhang, M. Experiment and mechanism study on cavitation erosion of ultrasound radiator in aluminum melt. *J. Huazhong Univ. Sci. Technol. (Nat. Sci. Ed.)* **2015**, *43*, 85–88.

26. Dong, F.; Li, X.Q.; Zhang, L.H.; Ma, L.Y.; Li, R.Q. Experiment and mechanism study on cavitation erosion of ultrasound radiator in aluminum melt. *Ultrason. Sonochem.* **2016**, *31*, 150–156. [CrossRef] [PubMed]

27. Liang, G.; Shi, C.; Zhou, Y.J.; Mao, D.H. Effect of ultrasonic treatment on the solidification microstructure of die-cast 35crmo steel. *Metals* **2016**, *6*, 260. [CrossRef]

28. Guo, C.X.; Li, X.Q.; Zhang, M.; Jiang, R.P. Dynamic characteristics study of ultrasonic transducer system for casting. *J. Mach. Des.* **2015**, *32*, 47–50.

29. Haghayeghi, R.; Ezzatneshan, E.; Bahai, H. Grain refinement of AA5754 aluminum alloy by ultrasonic cavitation: Experimental study and numerical simulation. *Met. Mater. Int.* **2014**, *21*, 109–117. [CrossRef]

30. Zheng, D.S.; Chen, R.R.; Wang, J.; Ma, T.F. Novel casting TiAl alloy with fine microstructure and excellent performance assisted by ultrasonic melt treatment. *Mater. Lett.* **2017**, *200*, 67–70.

31. Chen, D.D.; Zhang, H.T.; Wang, X.J.; Cui, J.Z. Investigation on microsegregation of Al-4.5%Cu alloy produced by low frequency electromagnetic casting. *Acta Metall. Sin.* **2011**, *47*, 185–190.

MDPI

St. Alban-Anlage 66

4052 Basel

Switzerland

Tel. +41 61 683 77 34

Fax +41 61 302 89 18

www.mdpi.com

Materials Editorial Office

E-mail: materials@mdpi.com

www.mdpi.com/journal/materials